DATE DUE

APR 0 6 2015	

designer
GENES

designer GENES

A NEW ERA
in the
EVOLUTION OF MAN

Steven Potter, Ph.D.

Random House / New York

Published in the United States by Random House, an imprint of
The Random House Publishing Group, a division of
Random House, Inc., New York.

RANDOM HOUSE and colophon are registered trademarks
of Random House, Inc.

Library of Congress Cataloging-in-Publication Data

Potter, Steven.
Designer genes: a new era in the evolution of man/Steven Potter.
p. cm.
ISBN 978-1-4000-6905-7
eBook ISBN 978-1-5883-6998-7
1. Genes—Popular works. 2. Human genetics—Popular works. 3. Genetic
engineering—Popular works. I. Title.
QH447.P68 2010
611'.01816—dc22   2009041659

Printed in the United States of America on acid-free paper

www.atrandom.com

2 4 6 8 9 7 5 3 1

FIRST EDITION

Diagrams by Jan Warren

To Stan, Leora, and Sue, for their support,

to my agent, Peter Riva, for championing my cause,

*and to my editor, Bob Loomis, for his unending patience
and insight*

CONTENTS

designer
GENES

PREFACE

Where did we come from? Where are we going? What did people look like in the past, and could they change in the future? These are questions that naturally pique our curiosity.

In searching for the answers we can examine the fossils from the earth. The first fossil record of the human lineage, separated from the apes, was found shortly after Darwin's death. This preman was named Homo erectus and lived over a million years ago. During the past century more than a thousand additional fossilized bones of human ancestors have been found, revealing a slow progression during the past seven million years to what people are today. In particular the skulls show a dramatic increase in size that started about two million years ago, eventually resulting in a tripling of the mass of the brain, from about one pound up to the current three. Instead of exceptional strength or speed this new creature seemed to rely on a vastly superior intelligence for survival. While the fossil record remains incomplete, the buried bones lead us to the inescapable conclusion that the ancestors of humans looked quite apelike, and very different from what we are today.

To better understand our past we can also look at the historical archive of our DNA, the substance of our genes. We have learned that DNA is not immutable, but rather it also changes slowly with time. It is a molecular clock. One can study the relationships of the various forms of life on our planet by examining their DNA sequences. It turns out that there is a very striking commonality in the genetic endowment of the various species. We all have surprisingly similar sets of genes. Everything seems related to everything, to some degree. Even the lowly fruit fly has some genes that are extremely similar to their counterparts in humans. Indeed, in a few cases it has been possible to swap a human gene into the fly, where it functions just fine. The fly doesn't seem to mind. And when one compares the DNA sequences of more closely related species the similarities absolutely overwhelm the differences. For example, the DNA of the human and the chimpanzee is approximately 99 percent identical. The molecular clock therefore agrees with the fossil record, indicating a close relationship between apes and man, and telling of the existence of a common ancestor millions of years in the past.

Evolution is very slow. Within the life span of a person evolution is practically invisible. There is nothing to see. It generally takes hundreds of thousands, or millions, of years for major changes to take place. The genius of Darwin was to recognize that even extremely small modifications that occur during a single generation can accumulate over long periods of time to give important consequences. Darwin noted in his book *On the Origin of Species,* "As many more individuals of each species are born than can possibly survive; and as, consequently, there is a frequently recurring struggle for existence, it follows that any being, if it vary however slightly in any manner profitable to itself, under the complex and sometimes varying conditions of life, will have a better chance of surviving, and thus be *naturally selected*" (italics added). Today we often summarize this statement as "the survival of the fittest." But Darwin's theory of evolution requires more. Not only must the fittest survive, but they must also pass their better qualities to their offspring. At around this same time the Austrian monk Gregor Mendel published a paper, "Experiments in Plant Hybridization," describing the transmissible nature of characters, such as flower color and stem length, in the garden pea. Although Darwin was unaware of Mendel's work he fully appreciated the heritable nature of traits. He stated, "From the strong principle of inheritance, any selected variety will tend to propagate its new and modified form." So the fittest will survive, and their offspring will also be more fit. And the cheetah keeps getting faster, to capture more prey, and the giraffe keeps getting a longer neck, to reach more vegetation.

While humanity's past is a series of facts, with some known and some waiting to be discovered, our future is more a matter of conjecture, and controversy. Some argue that the technology of civilization has removed the selective pressures of the past and that people have therefore stopped evolving. Others suggest that some selective survival and procreation do indeed remain today and that humans continue to evolve, albeit at a snail's pace. Yet another view is that we are about to enter a new era, where all of the forces that drive our evolution are about to change.

1

THE TRUE STORY OF ADAM

Both Lisa and Jack Nash were carriers for a Fanconi anemia gene mutation. They didn't know it, because they were quite healthy, but they both had one good and one bad (nonfunctional) copy of this gene. Their first child, Molly, unfortunately received two bad copies of the gene, one from each parent. As a result, she had Fanconi anemia, which is a disease with several manifestations, but the most lethal is a blood disorder: the body fails to produce enough blood cells.

Molly was born on July 4, 1994, and it was clear from the outset that things weren't right. As Lisa held her newborn daughter in her arms she knew there was a problem. Instead of a forceful cry there was but a whimper. And her thumbs were missing! Lisa quickly asked for a copy of a book, David Smith's *Recognizable Patterns of Human Malformation*. She had worked for years in a hospital as a nurse for newborns and knew that this was the standard reference book for birth defects. It lists diseases according to symptoms. Using this book Lisa was the first to diagnose her daughter with the very rare Fanconi anemia. Molly suffered from a severe case that would end her life in a few years unless a transplant could be performed. Donor bone marrow from an adult or an umbilical cord from a newborn would have blood stem cells capable of restoring her ability to make blood. But the donor cells must be well matched to those of Molly. Otherwise the donor cells would probably recognize Molly's cells as foreign, like bacteria, and launch a lethal rejection of her body. The transplanted blood cells, meant to save her, would then actually kill her. And despite an extensive search no compatible donor could be found.

Desperate to save their daughter, Lisa and Jack considered their options. If they risked going ahead with a bone-marrow transplant from a nonrelative donor that wasn't well matched, the chances of success were slim, below one in five. Doing nothing would result in Molly's certain death. At this point there didn't seem to be any other options. But then, as they thought more, they decided perhaps there was another choice. If they had another child, they might be lucky, and it might not suffer from Fanconi anemia (only 25 percent of their children would be unfortunate enough to receive two bad copies of the gene). But the odds were still poor that any additional child would provide a compatible transplant match for Molly.

They decided to go with a modified version of this option, very radical at the time. Indeed, they would be the first. They would take chance out of the equation. They would definitely have another child, but they would use the latest scientific advances to be certain that this child would not have Fanconi anemia, and that it would be compatible with Molly, and would therefore be able to save Molly's life.

The procedure was a modern-day variant of in vitro fertilization (IVF), which has been around for decades. Louise Brown, born in 1978, was the first child conceived through IVF. The procedure is remarkably simple in concept—eggs are mixed with sperm in a test tube, thereby achieving fertilization. The fertilized eggs, or zygotes, are grown for a brief period in the laboratory, and then surgically inserted into the mother, where they implant themselves into the wall of the uterus and develop into normal babies. This procedure has been enormously successful in helping otherwise infertile couples conceive. Approximately 1 percent of all babies now born in the United States are the result of IVF.

For the Nash family, however, another step was required, to make sure that the baby carried the correct combination of genes to provide a good transplant match with Molly. Eggs from Lisa Nash were mixed with sperm from Jack Nash, using normal IVF procedures to make fifteen early embryos. Three days after fertilization, when the embryos were at the eight-cell stage, a single cell was removed from each and used for genetic diagnostics, which didn't harm the embryos at all. The cells were analyzed for mutation of the Fanconi anemia gene, and for transplant match determination. An embryo with the correct gene combination was then transferred into the uterus of Lisa Nash. The result was the birth of a healthy boy whose umbilical cord blood stem cells provided a transplant that saved Molly's life. They had cured their precious daughter! And in the process they had acquired a healthy son, of course to be loved and cherished as well.

In a very prescient decision, the Nashes decided to name their new child Adam. The name acknowledges that Adam Nash, like the biblical Adam, represents the first of a new breed. Preimplantation genetic diagnosis, or PGD, is now performed routinely at many centers around the world. Single cells are removed from early embryos to determine their genetic makeup. It is almost exclusively used, at present, for the identification of embryos that are free of genetic disease carried by their parents. These centers all strongly insist that they are not involved in the production of designer babies, but rather the generation of healthy babies, lacking a deadly gene combination that would otherwise doom them to disease.

But the principle is established. The methodology of creating a batch of embryos and applying a genetic screen to determine which will be used is now entrenched. Currently we focus on the absence of gene variants known to cause disease. But we are on the verge of an incredible explosion of understanding of the functions of different forms of genes. In the future we will be able to completely sequence the DNA of each embryo, and to see what version of each gene is present. It will then be possible to add a large number of factors to the selection formula. Instead of just looking for absence of the Fanconi anemia gene, for example, it will be possible to choose on the basis of intelligence, musical talent, height, body build, mental health, eye color, hair color, and a host of other characteristics.

But this strategy of testing a small set of embryos—approximately ten to twenty are usually produced—is limited in its potential, because the desired gene combinations might not be found. Two new technologies offer even more sweeping possibilities. First, developments in the stem-cell field could make it possible to generate thousands of embryos to screen, instead of just ten to twenty. The ideal gene mixture will be much more likely to occur in a large group of embryos than in a small one. The second technology goes a giant step further, allowing one to take a single embryo and to modify its genes at will. The Nobel Prize in Medicine in 2007 was awarded to Mario Capecchi, Sir Martin Evans, and Oliver Smithies for the research leading to this breakthrough, which is routinely used today in research laboratories for the genetic modification of mice.

Our children are our biggest investment. They are what remain of us in the future. We want them to be the best that they can be. Our desires and our technologies have combined to place us on the proverbial slippery slope. It is not clear that we can change course now. The timing of the travel is subject to debate. Will it be five, ten, or fifty years? But the path we are following is apparent. And what is the final destination? Where is humanity headed?

2

FIGURING OUT WHICH GENES DO WHAT

Before we can make children with chosen sets of gene combinations to produce desired characteristics, we first have to figure out which genes do what. Right now we don't know. We are relatively ignorant of the gene type blends that would make a person smarter, stronger, and healthier. But, once again, there is a technological revolution under way that will rapidly change this.

To appreciate the revolution we must first review some underlying principles. This is a complex area that molecular biologists have been working for many decades to better understand. Nevertheless, the basics are surprisingly straightforward.

So, what are genes anyway? What do they do? You might think of genes as little computer programs, each designed to have a specific function. Maybe one gene has a contraction function and contributes to muscle contraction, while another might help the neurons of the brain communicate with one another. And just like a computer, which can have many programs but might only have a few in use at a given time, a cell has many genes, but only a fraction will be active at once. Every cell in your body has the same set of genes, but different cell types use different combinations of genes. And even a single cell will change the set of genes it is using with time, according to need.

How do genes do their work? Actually, genes don't really do the work themselves, they tell others to do it. The genes are just the information storehouse. Genes are made of DNA, and the DNA consists of strings of bases, called A, T, G, and C in molecular biology shorthand. James Watson, Francis Crick, and Maurice Wilkins shared a Nobel Prize in 1962 for their work defining the elegant double-helix structure of DNA. Just as a computer uses a binary code of ones and zeros, the human cell uses a quaternary code of these four bases. The sequence of the bases of the DNA provides the information. The DNA is like the hard drive.

But how does this information get used? As shown in the diagram on the next page, DNA is used to make ribonucleic acid (RNA), which is used to make proteins. The proteins, then, are the workers. Proteins have many different functions. Proteins in muscles allow the muscles to contract. Enzymes are proteins, and among other things, they catalyze the biochemical reactions that turn food into energy. Proteins in the surface membranes of the cell are like little eyes and ears, and allow the cell to sense its surroundings. Some proteins are secreted from a cell and float away to nearby cells, communicating with them. The messages received at the outer surface of the cell must be carried to the DNA-containing nucleus, again by proteins, where the cell can

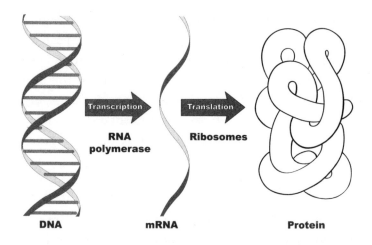

DNA   **mRNA**   **Protein**

make appropriate changes in the way it activates its genes in response. Indeed, proteins do almost everything in the cell that needs to be done.

It is interesting to note that not only are proteins the products of genes, but they also facilitate the process of gene expression. The protein RNA polymerase makes the RNA copy of DNA. This is called transcription, because it is just copying, or transcribing, the base sequence of the DNA into the base sequence of the RNA. The resulting RNA, called messenger RNA, or mRNA for short, serves a courier function, carrying the sequence information from the nucleus, where the DNA resides, out to the cytoplasm of the cell, where protein is made. In the cytoplasm there are machines, called ribosomes, made up of many dozens of proteins, which have the job of translating the base sequence arriving in the mRNA into the subunit sequence of newly synthesized protein. There is circularity in the process, with many proteins necessary to activate, or express, a gene, which in turn makes more protein.

Many of the molecules of the body are polymers, made

by joining a large number of building blocks together. DNA, as mentioned earlier, is a double helix made up of the four bases A, T, G, and C. The closely related RNA also has four bases, A, U, G, and C, with a U in place of the T in DNA. And proteins are made up of subunits called amino acids, which have different physical properties that give them different functions. Everyone has heard of protein as a part of the food we eat. When we digest this protein we break it down into the subunits, the amino acids, and then our cells use these amino-acid building blocks to synthesize their own proteins.

There are twenty different amino acids, almost the same number as there are different letters in our alphabet, and each protein is on average a few hundred amino acids long. This allows an enormous variety of proteins to be made. Imagine how many different words you could make if each word was a few hundred letters long, and if the change of even a single letter in the word could dramatically change its meaning. This is exactly the situation for proteins. In many cases the change of even a single amino acid can make the difference between life and death.

An important principle in molecular biology is the complementarity of bases. Watson and Crick showed that in double-stranded DNA the base A is always opposite the base T, and the bases G and C are also always found across from each other. It turns out that an A "fits" next to a T, and loosely binds with it, and the same is true for the base pair G and C. When DNA is replicated the two strands are pulled apart, and each strand is used to make a complementary strand, restoring the double stranded structure and giving two DNA copies in place of the original one. Proteins, including DNA polymerase, also carry out this DNA replication, using the principle of base complementarity to make sure that the newly synthesized strands of DNA have the correct sequence. Again, an A in one strand is always placed next to a T in the other, and a G next to a C. A single round of DNA replication must take place before a cell divides, so the two daughter cells can each receive a complete copy of the DNA.

RNA polymerase also uses this principle of base complementarity when it takes one of the strands of DNA and transcribes an RNA copy from it. But, as we've mentioned, RNA is not exactly the same as DNA. When RNA is being made the RNA base U will go next to the DNA base A. The diagram below shows one strand of DNA, called the coding strand because it encodes protein, being transcribed into RNA.

You might wonder how the base sequence of DNA is used to determine the amino-acid sequence of protein. There are only four bases, but twenty amino acids, so clearly there is not a one-to-one correspondence. Cracking the genetic code was one of the major early accomplishments of molecular biologists. It took many years and a great deal of ingenious experimentation to understand this process. It turns out that three bases are needed to encode, or specify, each amino acid. Each base triplet encoding a single amino acid is called a codon. For example, the codon ATG encodes the amino acid methionine, abbreviated "met." Another three bases will encode a different amino acid, and so on. A gene, therefore, consists of DNA, which is a chain of bases, with each sequential codon of three bases specifying a particular amino acid of the encoded protein. Different genes have different base sequences, and therefore encode different proteins with different roles.

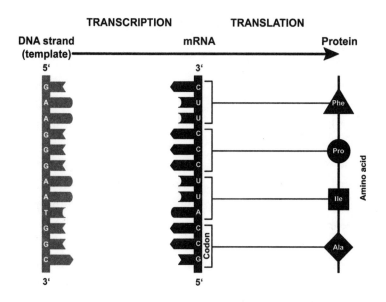

As noted previously, all of the cells of the body have pre-cisely the same set of genes. We all started as a single cell, the fertilized egg, and before that cell divided, its DNA was duplicated, so that each daughter cell received the same set of genes. Complete DNA replication also preceded each subsequent cell division, resulting in all of the cells of the body having the same DNA. The brain and liver cells, al-though they have the same genes, are different from each other because they use, or activate, different sets of genes.

THE ORIGIN OF LIFE

The complexity of gene expression raises an interesting question. How did life begin? Viruses, bacteria, plants, and people must all use their genes to survive. Yet activating a gene requires the complicated interconnected interactions of a large number of molecules, including DNA, RNA, and a host of proteins. Scientists who study the origin of life itself face the very difficult problem of figuring out how this all started. How did the very first life form get all of these pieces of the puzzle together at once to allow gene expression? Could this have happened by chance, in the so-called primordial soup? This is a field where there are still more questions than answers, but some exciting progress has been made. Sidney Altman and Thomas Cech shared a Nobel Prize in 1989 for discovering that RNA molecules could on occasion act like proteins. Certain RNA molecules have enzymatic activity, like some proteins. These RNA enzymes are called ribozymes. Further, it is known that some viruses have genes made of RNA, and not DNA. It therefore appears that a single type of molecule, RNA, might be capable of fulfilling all of the functions of DNA, RNA, and protein, thereby greatly simplifying the problem of the origin of life. Maybe in the beginning there only needed to be one type of molecule, RNA. Nevertheless, we are still far from understanding how that first spark of life emerged.

GENE DIFFERENCES BETWEEN PEOPLE

While all people in general have the same complement of genes, the genes are not identical in sequence from person to person. When a gene is mutant, or different from normal in its base sequence, the result can be a completely inactive gene, which sometimes causes a genetic disease. For example, a deletion of a block of the DNA sequence would result in a corresponding deletion of a block of the encoded protein, which would likely destroy the protein's function. In addition, at the end of each gene there is a stop codon, which tells the protein-making machinery where to terminate protein synthesis. If a mutation results in the appearance of a stop codon in the middle of the gene, instead of at the end where it belongs, then a truncated protein will be made, again likely inactive. Another kind of mutation that usually completely inactivates a gene is called a frameshift. The triplet codons are read in a very precise way, one right after another, one set of three bases after another, and if a single base of DNA, for example, is deleted, then the whole reading frame is disrupted. The machinery keeps reading three bases at a time, but now they are the wrong three bases—a base that was supposed to be read as the third base of a codon is now the second base, because the second base was deleted. And the first base of the following codon becomes the third base of the codon with the deletion. Everything is out of order, and the result is usually very destructive, causing the gene to encode a completely incorrect amino-acid sequence after the frameshift mutation.

We all carry some genes with such inactivating mutations, but since we have two copies of every gene, one from each parent, and since for most genes we only need one good copy, this isn't usually a major problem. For example, more than one person in a hundred has a mutant CFTR gene, but also one normal copy of the gene. These people are perfectly healthy. Loss of both functional copies of a gene, however, can have devastating consequences. Approximately one newborn in three thousand will have inactivating mutations in both copies of the CFTR gene, resulting in cystic fibrosis. The CFTR gene encodes a protein that transports molecules across the outer membrane of the cell. Without a working copy of this gene, lung function is impaired and life expectancy dramatically shortened.

Genes can also have single base substitutions, such as an

A changed to a T, which alters a single codon. Perhaps an AGU codon will be changed to a TGU. But there are sixty-four different codons, and only twenty different amino acids, so on average there are about three different codons that specify each amino acid. This means that sometimes a changed codon will still encode the same amino acid. Indeed, natural selection has evolved a genetic code with all of the codons for a single amino acid having a very similar sequence, making it more likely that single base changes will not change the amino-acid sequences of encoded proteins. Nevertheless, often they do. And when there is a codon difference in the DNA that causes an amino-acid difference in the protein it can sometimes have devastating consequences, if it is a key amino acid. In other cases the result is subtle, with a small change in the function of the resulting protein.

These single base differences in the DNA of individuals in a population are called single nucleotide polymorphisms (differences), or SNPs. It is clear from studies already carried out that there are millions of SNPs (pronounced "snips") in the human population. Some of these occur in the codon regions of genes, perhaps altering the amino-acid sequences of encoded proteins, and others are found in noncoding parts of the DNA, perhaps disturbing the regulation of gene-activation patterns. These SNPs presumably account for much of the genetic diversity of features that we see among people. At present we are, with very few exceptions, unable to associate specific SNPs with particular features in a person. But again, that is about to change.

3

SEQUENCING THE
HUMAN GENOME

Before we can relate rare differences in the letterings of genes to the physical characteristics of people, we have to establish some of the basics. To start, how many genes do people actually have, and what are their "normal" sequences? By the mid-1980s it had become possible to determine DNA sequence, hundreds of bases at a time, by using two Nobel Prize–winning approaches developed by Frederick Sanger (who had previously won another Nobel Prize for work in protein sequencing), and Allan Maxam and Walter Gilbert. The procedures were laborious and somewhat tricky, and the quality of results varied from laboratory to laboratory, but these new techniques represented a great leap forward from previous approaches that required an entire lab to work for years to determine even a hundred bases of sequence. Many of the research papers in the 1980s, from labs scattered around the world, were dedicated to the determination of the DNA sequence of a single gene. It was realized, however, that this piecemeal approach was inefficient. There would be great economy of scale achieved by creating a few sequencing centers, which would become masters of DNA sequencing and would determine DNA sequence on an industrial scale.

The Human Genome Project, funded by the U.S. government, was formally launched in October 1990 and completed thirteen years later, in 2003, with a total cost of a little under four billion dollars. The goal was to determine the entire DNA sequence of the human genome. There were major contributions from the private sector and from other countries, but clearly the driving force behind this endeavor was the U.S. Human Genome Project.

The Human Genome Project has been compared to the moon-landing project in importance. They both represent monumental achievements of mankind. The moon landings, in historical perspective, represent the first extraterrestrial step toward mankind's exploration of space. The Human Genome Project, however, will likely in the end have an even greater impact on our lives.

THE AMAZING HUMAN GENOME

In one sense the Human Genome Project represents a remarkable instance of a genome, through the people that it encodes, completely defining itself. The DNA defined the people that in turn defined the DNA. The results led to a much deeper understanding of the way the human organism works. Effects will range from the more complete definition of genetic-based disease to understanding developmental biology and birth defects (how the human forms from the fertilized egg) to the use of stem cells to regenerate and repair damaged organs, and to the ultimate self-directed modification of this now-detailed genome, to create novel versions of the species Homo sapiens. This is indeed a remarkable genome, capable of first defining and then refining itself.

It is truly amazing to consider that the human genome has not only completely characterized itself, determining the exact sequence of its three billion bases, but it now faces a future in which it can drive its own modification and evolution. As we will see later, the technology exists right now to allow precise alteration of the genome. In a relatively short time we will be able to choose the genes of the next generation. It is possible to foresee a future when our species, already quite smart, gets more and more intelligent, acquiring with time increased capacity for self-improvement. It is interesting to consider the consequences of a never-ending cycle of increasingly intelligent modification of the genome. It is like a chain reaction, a thermonuclear explosion in the evolution of the human species.

NUMBER OF GENES

So, what are some of the specific revelations to come from the Human Genome Project? The first, and perhaps most shocking, is the surprisingly small number of genes it takes to make a person. Before the completion of the project, scientists had generally overestimated the number of genes human DNA would include. We knew the total amount of DNA present per cell, the total number of bases that were there, and we knew about how much DNA it took to encode an average protein, or how many bases an average gene would need. We also knew that a significant part of the

DNA was not genes, but something else, an issue we will return to later. When we crunched the numbers we came up with a ballpark total of one hundred thousand genes.

One of the challenges facing us after determining the human DNA sequence was figuring out where the genes were. We now had a long string of base sequence, three billion bases total, but where were the genes? Molecular-biology techniques allow us to identify regions of the DNA that are copied into RNA, which is the first step in the activation of a gene. Using this tool, and some others, yielded our current estimate of about twenty-five thousand, or fourfold fewer genes than expected. This number is gradually creeping up, as more are discovered, but it appears likely that in the end the final number will remain under thirty thousand.

It is rather shocking that it only takes about twenty-five thousand genes to make a person. Think of it. That first cell, the fertilized egg, a single microscopic speck of protoplasm, only has twenty-five thousand genes to direct its development into a complete person. This is an extraordinary transformation. People are incredibly complicated. They have hundreds of very different cell types, many complex organs, including a brain with trillions of neurons, with thousands of trillions of intricate neuron-to-neuron connections. How can a little cell equipped with only twenty-five thousand genes make all of that? It hardly seems possible. This is another interesting topic we'll explore later.

DNA SEQUENCE COMPARISONS ACROSS SPECIES

Scientists have now also determined the DNA sequences of the genomes of many other organisms, with shocking results again. For example, consider the lowly fruit fly, zipping around your ripe banana. The genome of this tiny little beast has about fourteen thousand genes. This is about half as many as a person! If it takes only twenty-five thousand genes to make a person you might think it would take a very much smaller number to make a fly. But the total gene numbers required to encode people and flies are not hugely different.

Even more humbling, we found when we sequenced the DNA of a plant, a little weed with the scientific name *Arabidopsis,* that it also has twenty-five thousand genes, about the same number as us! Now, this weed is a far simpler organism than a human, yet it has just as many genes. It is clear that organism complexity and genetic complexity don't always show a linear relationship.

This comparison of DNA sequences becomes even more intriguing as we look at species more closely related to humans. Consider the mouse, which, like a person, is a mammal. It turns out that mice have almost exactly the same number of genes as people. Indeed, this is true for all mammals, including cows, pigs, rats, elephants, whales, and lions. They all have about twenty-five thousand genes.

Furthermore, the close genetic relationship among mammals goes deeper than total gene count. The Nobel laureate Mario Capecchi has stated that mouse and man are 99 percent genetically identical. This doesn't mean their DNA sequences are 99 percent identical, because they aren't, but it means that if you take a human gene and look in the mouse genome there is a 99 percent chance that you will find a corresponding gene that is extremely similar. That is, mouse and man have very much the same set of genes. Then why do we look so different? Because there are lettering, or coding, differences between the mouse and human gene counterparts. It turns out that all mammals are equipped with the same basic package of genes, but each gene varies somewhat in its exact base sequence from one species to the next.

And what about the chimpanzee? Of course, as a fellow mammal, its DNA encodes essentially the same number of genes as for humans. But in this case there is, perhaps to be expected, an even more striking similarity in the DNA sequences. Out of the three billion total bases there were only thirty-five million sequence differences, or a little over 1 percent, between chimp and man. Indeed, for one-third of their genes the two species encode exactly identical proteins. The sequence differences between chimp and man were only about ten times greater than those between one person and another. This striking similarity between the chimp and human genomes clearly substantiates the very close evolutionary relationship between these two species.

An interesting pattern emerges when we consider the many different mammalian DNA sequences that have been determined. First, while all mammals have the same basic set of genes, their precise base sequences are not identical. If you compare two different individuals of one species, like two different people, then you find about three million base-sequence differences, out of the three billion total. These three million base-sequence differences are what make you different from your neighbor. And if you compare the DNA sequences of two closely related species, like chimp and human, then you find about ten times more sequence differences. According to this DNA sequence measure, a chimp is only about ten times more different from you than your neighbor is. If you don't like your neighbor then you might point this out to him. And if you compare the DNA sequences of more distantly related mammals then of course you will find still more DNA sequence differences.

Another way to view this is to consider the human genetic endowment as a string of DNA, three billion bases long, with all of the DNA of all the chromosomes lined up end to end. The genetic code that defines you would then be a single line of bases, such as ATGCCTTAGCCGA . . . for three billion bases. In comparing your DNA to that of your neighbor, both sequences would be identical, on average, for a thousand bases before you would find a single base difference, such as a T in place of a C. Then about another thousand identical bases would go by before another single base difference. This 99.9-percent identity is quite striking at the DNA-sequence level, even though you might not consider yourself all that similar to your neighbor. And your DNA compared to that of a chimp would again show long identical stretches, on average a hundred bases long, before finding single base differences.

Although chimp and man are a lot alike at the DNA-sequence level, there are clearly some pretty significant differences between the two species. Can we look at the two DNA sequences and begin to establish which versions of which genes account for the trait differences? Could we, for example, determine that humans are smarter because of a certain distinct version of a specific gene? Much effort is being invested in this area, with some limited progress. For example, there exists a British family suffering a severe speech deficit, which has been traced to a sequence difference in a single gene, named *FOXP2*. The deficit is called verbal dyspraxia; the affected family members are unable to produce the coordinated muscular movements needed for speech. Interestingly, chimps also carry a modified version of this gene. This strongly suggests that differences in speech ability between chimps and humans might be related to sequence differences in this gene.

But in general, it is extremely difficult to relate the sequence DNA differences between chimp and man to trait differences. Chimps and people have a large number of trait differences, including arm length, hairiness, intelligence, shape of the cranium, and so on for an extremely long list depending on the level of detail. And the chimp has a very large number, thirty-five million, of DNA-sequence differences from the human. While one-third of genes are identical, two-thirds are not. So, if you have perhaps a thousand trait differences, and fifteen thousand gene differences, it is almost impossible to decipher which gene versions are responsible for which trait differences. The FOXP2 example represents a very rare example where the same gene-sequence difference seen in the chimp has been found in a human family, with a trait difference also resembling that observed in the chimp. It is unlikely that we will come across many examples of such a fortuitous coincidence.

What we would really like to have are the DNA sequences of a large number of people. Different people would have smaller numbers of trait differences, and smaller numbers of DNA-sequence differences, making it much easier to figure out which versions of which genes are responsible for which traits. Although most people are unaware of it, this sequencing of the DNA of large numbers of people is already under way.

4

THE SEQUENCING REVOLUTION

I t is possible that when historians look back at the current century they will refer to it as the Age of Sequencing. Powerful new technologies are rapidly emerging. There are gene chips, resembling computer chips in their construction, which can sequence a million bases overnight. At one point this appeared to be the winning technology, because gene chips kept getting less expensive and more powerful each year. But then other technologies leapfrogged gene chips, allowing scientists to determine far more DNA sequence both faster and cheaper. These approaches use either tiny beads or microscopic DNA spots on a slide to carry out millions of sequencing reactions at once. In the world of the molecular biologist these methods are referred to as next generation, or NexGen, sequencing. Several companies are even developing systems that will eventually make it possible to sequence fantastically large numbers of single DNA molecules at once.

The end result is that, as of this writing, it is possible to use one of these NexGen sequencing machines to determine 200 billion bases of DNA sequence in a single run. Remember, a human has approximately three billion bases of sequence. It might seem, therefore, that one run of the machine, requiring only a week, would determine the sequence of fifty people or more. But in fact we need what is called high coverage—we need to determine the sequence many times over, to make sure that we can fit together all of the pieces of sequence that are generated, to make sure there are no gaps, and to make sure that we sequence both copies of every gene we carry. But even with these reservations it is now possible with this new technology to sequence a person's DNA in just one week at a cost of about ten thousand dollars. This is a most spectacular advance, considering that the Human Genome Project spanned thirteen years and consumed almost four billion dollars to determine the first single human DNA sequence. The price of the genome sequence has come down dramatically.

And the price is going to continue to plummet as these technologies improve and compete for a vast and lucrative market. It has been predicted by many experts in the field that in the near future it will be possible to determine a person's complete DNA sequence in a couple of days, or less, for a cost of only a thousand dollars. At this point DNA sequencing will become a routine diagnostic procedure. The doctor will want to know what your DNA says about your disease tendencies. Every newborn baby will have its DNA sequenced, assuming it wasn't sequenced before the baby was born. We will amass an enormous body of sequence data from thousands, and eventually millions, of people, along with medical records that will allow us to make that sought-after correlation between DNA sequence and attributes.

THE 1000 GENOMES PROJECT

An example of how this new technology for sequencing multiple human genomes is moving forward is the 1000 Genomes Project, which was launched in January 2008, with considerably less fanfare than the Human Genome Project, but likely in the long run with even more of an impact. An international consortium of centers in England, China, and the United States is carrying out the sequencing work. Human DNA sequence will be churned out at a rate of over eight billion bases a day, with over six trillion bases of sequence generated in total. And while the original Human Genome Project cost about four billion dollars, the 1000 Genomes Project is estimated to cost only a hundredth of that, about forty million dollars.

The purpose of the 1000 Genomes Project is to sequence the specific genetic variation present in the DNA of a thousand different people. The goal is to identify gene combinations that contribute to disease. While many genetic diseases are the result of a severe, devastating mutation of a single gene, it now appears likely that many more diseases are caused by more modest modifications of a number of different genes. The altered genes still work, just not quite as well as normal. Each variant gene is a single contributing factor, but it takes an accumulation of several, or dozens, or perhaps hundreds of these contributing factors to result in disease.

For diseases where a mutation of a single gene is the underlying cause we have developed procedures that can routinely identify the responsible gene. Scientists study families with affected individuals, not sequencing their entire genomes, but simply looking at the patterns of selected single nucleotide polymorphisms (SNPs), or differences, present in different individuals. Perhaps a million of these SNPs will be defined for each person, with each single-base polymorphism marking a small region of the genome. All of the diseased individuals of the family will carry the mutant form of the responsible gene and the associated specific SNP for that region of the chromosome. By finding the regional SNPs common to diseased family members one identifies the responsible region of the DNA, and the known genes in that region can then be sequenced in normal and diseased individuals. The sequence comparisons serve to identify the single mutant gene whose altered sequence causes the disease. This sort of strategy has worked for a large number of genetic diseases. The Online Mendelian Inheritance in Man website (www.ncbi.nlm.nih.gov/omim) lists a few thousand gene-map positions with associated diseases and conditions.

But the situation is much more complicated when the inheritance pattern is multigenic, with many contributing genes. And a multitude of diseases, including schizophrenia, autism, and many forms of heart disease and cancer appear to have a multigenic cause—variant forms of many genes can contribute to the disease, and if a sufficient number of genetic contributing factors are present then the disease is likely to occur. For example, there might be thirty or more altered genes that can add risk for a disease, and if you have any five of these then your chances of having the disease are very high. It is clear that the resulting inheritance pattern is extremely complex, as two diseased individuals could, for example, have variant forms of completely different sets of five risk genes. And some genes might contribute more than others, and some variant gene forms might make more important contributions, making the situation even more complex.

For these conditions with a compound genetic basis the SNP mapping strategy is often inadequate. The superficial look at the DNA that it provides, examining only a million bases of the three billion present, just doesn't give enough information. The critical sequence differences remain hidden in the remainder of the genome.

The 1000 Genomes Project begins to attack this problem. The initial goal is simply to begin to define the genetic variation present in the population. How many different forms of different genes are present? The next step will be to relate the DNA sequences to medical histories. For example, about one percent of the population suffers from schizophrenia. One would predict, therefore, that about ten people from the set of a thousand being sequenced would have schizophrenia. It is known that this disease has a very important genetic contribution, although environment also plays a role. We could look at their DNA sequences and see if schizophrenics share any recognizable patterns of altered genes. Depending on how genetically complex the underlying basis of schizophrenia is, these results could begin to define which forms of which genes represent risk factors. The identification of the responsible genes could lead to a much deeper understanding of the disease, which in turn could result in improved treatments. And, of course, if we know which versions of which genes can result in schizophrenia, then we could choose to avoid passing these genes on to our children.

The more individual genomes we sequence the more powerful the analysis. It is remarkable that in the year 2008, only five years after the completion of the Human Genome Project, we began to analyze a thousand genomes, for only one percent of the cost of the earlier project. And the revolution in DNA-sequencing technology is far from over. Indeed, it seems to be accelerating. Later in the same year, 2008, yet another even more massive sequencing project was launched, with the goal of determining the sequences of about twenty-five thousand genomes. The purpose of this study, the International Genome Cancer Consortium, is to better understand the genetic differences that cause cancer. For example, five hundred patients with ovarian cancer will be selected, and the complete DNA sequences of their normal tissue, as well as tumor tissue, will be established. We already know that mutations in certain classes of genes, called oncogenes and tumor-suppressor genes, play important roles in driving the formation of cancers. But there are many genes in these classes, and in most cases we don't know what types of mutations in which genes are responsible for the various cancers. This enormous sequencing project will begin to provide the answers.

In the not too distant future we will have millions, and then tens and hundreds of millions, of DNA sequences available. Imagine the DNA sequences of a million individuals, with about ten thousand of them likely having schizophrenia. What gene variants will this set of sequences have in common? By identifying those DNA-sequence differences that are more frequent in the patients with schizophrenia we will precisely define the gene combinations that represent risk factors for this disease. This will allow a robust and exhaustive analysis of the genetic contributions to the disease. Even minor gene variants that pose only a slight increase in risk will be identified. And imagine a similar genetic analysis for heart disease, cancer, and dozens of other diseases.

The concept that analysis of DNA sequences of large numbers of individuals will allow the identification of gene variants responsible for disease is extremely important. It provides the foundation that allows us to, in similar fashion, determine the genetic combinations that define traits such as intelligence. The basic notion is very straightforward. People with specific disease tendencies, or physical traits, will share distinct sequence features in their DNA. These sequence variations define the gene versions responsible for the disease or trait.

In practice, however, the analysis is more complicated. In comparing the DNA sequences of any two individuals there are about three million base differences, and the vast majority of these are random noise, unrelated to a disease or trait of interest. So, if the DNA sequence of one person with schizophrenia, for example, was compared to that of a person without the disease there would be about three million differences, but it would likely not be apparent which of these differences contributed to schizophrenia. It wouldn't be completely hopeless, as one could focus the analysis on genes known to function in the brain, and perhaps interesting differences would emerge. But it is unlikely that such a sequence analysis of just two individuals' DNA would be very informative. Instead it will be necessary to sequence the DNA of a large number of individuals with the disease or trait of interest, and to compare these to the sequences from a large number of people without the disease or trait of interest. A statistical analysis then identifies variant genes that consistently associate with the disease, above the random noise.

Let us consider a hypothetical example to better illustrate the point. In the vernacular of Einstein this might be considered a gedanken, or thought, experiment. The simplest example of multigenic inheritance of a disease would be a situation where two genes are involved, and both must be altered for the disease to be present. Let us suppose that you have the DNA sequences of one hundred patients with the disease, and of one hundred normal people without the disease. In comparing the individual DNA sequences of the normal group to one another we would find many differences, but they would be randomly scattered about the genome. There would be no pattern to the variation. On the other hand, in comparing the sequences of the patients with the disease to those of the normal group there would be two hot spots, the genes responsible for the disease, where DNA-sequence differences would always appear. In this simple example each patient would always have a variant form of both genes, and it would be very easy to identify the responsible genes, even with a relatively low number of patients and normal DNA sequences.

In more complicated examples the same principle applies, but more DNA sequences would be required to identify the statistically significant hot spots, since a particular gene might be variant in only a fraction of individuals with a disease. But if there is indeed an underlying genetic basis, and if enough DNA sequences of diseased and normal individuals are defined, then it will be possible to determine which versions of which genes contribute to the disease.

Consider an extreme situation, where there are a thousand genes that can contribute to a disease, but the contribution of each gene variant is minor. It might take the combined presence of a few hundred to result in disease. This is clearly a very difficult problem to solve. Which genes are responsible? Different people with the disease would have overlapping sets of responsible genes. But, once again, the principles of analysis remain the same. First, we must work with complete genome sequences. We can't use simple "old fashioned" SNP data, which define only a fraction of the sequence variation present. Complete genome sequences will, of course, see all of the sequence differences present, and won't miss rare variants that might be responsible for the disease. Second, we must have complete sequence data for large numbers of individuals, both with the disease and without the disease. We are looking for trends and statistically significant differences, which can't be done effectively with sequences from small numbers of people. Third, and perhaps most important, the gene variants that contribute to the disease will be more common in people with the disease. Of course the people with the disease have more of them, that's why they have the disease. A statistical analysis of the sequence data will identify those gene variants that are more frequently found in diseased individuals, and therefore are likely causal. Perhaps, for example, a particular sequence version of a gene is found in 1 percent of people without the disease, but in 10 percent of people with the disease. If enough people had been examined to make this difference statistically sound, then this gene variant would be strongly implicated in contributing to the disease. Of course, the more complex the underlying genetic basis, the more genomic sequences required for the analysis. In some cases it might be necessary to sequence the DNA of millions of people. But as the price of sequencing continues to drop, a time will come when this won't be a problem.

And the same type of analysis will allow us to find gene combinations that result in desirable traits. Consider longevity. Some people are relatively hearty and healthy in their nineties, while others seem to be falling apart in their fifties. Certainly lifestyle is a major factor, but there is also a very strong genetic contribution to life span. In some families people just seem to live longer and healthier than in others. And research studies with model organisms, such as fruit flies and mice, have already found mutant forms of some genes that can dramatically extend or shorten life span. If we had the DNA sequences of a million individuals, and knew their life spans, it would be quite possible to find the gene combinations that give a long and healthy life. Another straightforward strategy would be to sequence the DNA of a hundred centenarians, to find the common features of their DNA sequences that contribute to their longevity.

Many other human characteristics, including weight, appearance, athletic ability, and intelligence also show multigenic inheritance. Again, the same studies that define the genetic basis of disease will provide the data that can be used to identify genes responsible for these various attributes. It is only a matter of time before we will have a remarkable understanding of the genetic basis of hundreds of human traits.

5

DOGS

Dogs provide an interesting example of man-driven evolution. Consider the astonishing variety of different dog breeds. There are tiny dogs like the Chihuahua, which weigh only about five pounds, and enormous dogs like the Saint Bernard, which can weigh over two hundred pounds. Some dogs are extremely intelligent, with the border collie, retriever, poodle, and German shepherd often placing at the top of rankings. Their intelligence is comparable to that of a human of about two and a half years of age. But they are more obedient than a toddler. These dogs generally recognize new commands with under five repetitions, and then obey first commands almost always, while other dogs, such as the basset hound, chow chow, Pekingese, and bulldog rank near the bottom in intelligence, often requiring a hundred or more repetitions to learn, and even then failing most of the time to obey the first command. There are many

hundreds of dog breeds, with a wide variety of tempera-
ments, appearances, and body types. There is such an incredible assortment of different dogs that it is easy to forget
that they are all the same species, *Canis lupus familiaris*.
This means, in essence, that even a Chihuahua and a Saint
Bernard (assuming that the obvious physical challenges
could be overcome) could mate and produce live and fertile
offspring.

Where do dogs come from? Charles Darwin suggested
that they might come from multiple sources, including
wolves, jackals, and coyotes, thus explaining in part their
diversity. Recent DNA-sequencing results, however, con-
firm that all dogs are in fact derived from the wolf, with no
input from any other species. This means, surprisingly, that
all of the diversity of dog types in the world today came
from a single source, the wolf.

How did it all begin? This is a matter of conjecture, but
it is easy to suppose that wild wolves might have been
found living near man around fifteen thousand years ago,
enjoying leftover bones and such. At first, natural selection
would drive some wolves to be more tolerant of the presence of humans, or tame, because this would allow greater
access to the human garbage food source. Then perhaps an
injured wolf, or an abandoned litter, was taken in and
nursed by people. The DNA evidence, which shows a
strong similarity for all dogs, suggests that there might have
been only a few such domestication events. This speeded
the process of evolving the wolf into the domestic dog.
These early wolf dogs would be subjected to what is called
artificial selection. In the wild there is natural selection at
work, with the strongest, smartest, and fastest surviving
better to make more wolves. But once under the care of hu-

mans, survival would be dictated by a new set of rules. Animals that liked to bite people, for example, probably did not last long. But dogs are natural hunters that could assist in the search for food. And humans would benefit from the early warning system of dogs barking during an approach from unwelcome visitors, including other humans or wild animals. So, people-friendly watchdogs, with their heightened senses of hearing and smell, would provide an asset to early humans.

The earliest discovered true dog skeletons date back around eight thousand years. The ancient Egyptians, from four thousand years ago, leave us written descriptions as well as drawings of a few distinct dog breeds, including a dog clearly bred for speed, much like the greyhound. The Greeks acquired these dogs and used them for hunting rabbits, deer, and boar. Other early breeds included the wolfhound, bred in Britain; one was given to a Roman consul in 391 B.C. as a gift.

People selected dogs for a variety of features, including hunting ability, companionship, intelligence, herding ability, protection, and looks. The wolfhound was bred for, among other things, swimming ability, and actually has webbed feet. Most of the over four hundred breeds that exist today were developed in the last 150 years. This demonstrates a remarkably rapid evolution of a great number of different dog breeds. The general strategy used to create a dog breed is very simple. One might start by crossing, or mating, dogs from two different existing breeds, to maximize genetic diversity in the offspring. Then there is systematic selection, choosing the pick of the litter, those animals showing desired characteristics, such as speed, friendliness, hunting ability, or being a good watchdog, and then interbreeding these selected animals to give the next generation, and so on. Continued brother-sister matings coupled with systematic selection results, in surprisingly few generations, in a new breed of dog with a new set of characteristics. The new breed is genetically pure, because the continued inbreeding eliminates genetic diversity. And the new dog can have a very distinctive set of features because of the artificial systematic selection for those very features. The Doberman pinscher, the Australian cattle dog, and the whippet were all developed in this manner. New breeds carry unique combinations of genes from the previous breeds used to derive them.

Some of the working breeds of dogs show extremely impressive levels of intelligence. For example, a Border collie named Rico learned the names of over two hundred objects, mostly toys, balls, and stuffed animals. Rico and his owner would be placed in one room, with many objects scattered on the floor of a different room. The owner would ask Rico to fetch a specific item, and Rico would come back with the right one about 95 percent of the time. Even more extraordinary, Rico was able to use reasoning to figure out the names of new objects. In this case an unknown object would be placed among things that Rico already recognized, and the owner would ask Rico to fetch it, using the new object name that Rico had never heard before. Rico was able to determine that the name did not correspond to any of the items that he already knew, so it had to have been the name of the unfamiliar object. Further, he was able to remember the new name for months. The report of Rico inspired a search for other dogs with exceptional intelligence, resulting in the finding of Betsy, another Border collie. Betsy scored even higher on intelligence tests than Rico, and had a vocabulary of over three hundred words. Juliane Kaminski, a scientist who has worked with both Rico and Betsy, said, "Dogs' understanding of human forms of communication is something new that has evolved. Maybe these collies are especially good at it because they're working dogs and highly motivated, and in their traditional herding jobs, they must listen very closely to their owners."

This dog story is an interesting demonstration of evolution at work. In an extremely short period of time, in evolutionary terms, the wolf evolved into the dog, including all the great variety of dog types we have today. This is one evolutionary event that was not only watched by man but, indeed, was directed by man. It is only one example of many domestic animals and plants that illustrate the incredible power of artificial selection. And remember that this process was not driven by any scientific understanding of the underlying genetics. We didn't know which gene did what, or which animal carried which DNA-sequence version of a particular gene. We simply took the animals that showed the most desired features and used them to breed the next generation. And even with this primitive approach we could create hundreds of diverse dog breeds in only 150 years.

It is also remarkable to consider that the wolf had enough genetic diversity within its genome to give rise to all the dog breeds we see today. Wolves all look pretty much alike, and you might think that if you continued to breed wolves you would keep getting more wolves. But just as different people have millions of DNA-sequence differences from one to the next, even a single wolf can carry within its genome considerable genetic diversity, since all mammals have two copies of each gene, which do not need to be identical to each other. The wolf experiment carried out by man during the past few thousand years shows how it is possible to separate out many different traits to produce many different breeds of dog, starting with just the wolf. This is what happens when we create offspring with different combinations of the gene variants present in the original parents.

Imagine how much more rapid and powerful this process of directed evolution would be if we actually knew which forms of which genes were responsible for specific characteristics, and if we could scan for the presence of these genes, and even directly change genes if necessary, to create the desired combinations in the following generation.

6

WHAT IS ALL OF THAT DNA DOING, ANYWAY?

We already discussed the prime function of DNA, to serve as a repository of information. The lettering code of its base sequence defines proteins, which are the workers that make muscles contract, make enzymes that digest food, synthesize molecules that allow brain cells to talk to one another, and so on. The central dogma of molecular biology is that when a gene is turned on, the sequence of that gene's DNA is first copied into messenger RNA, which then travels from the nucleus, or command center of the cell, out to the cytoplasm. There the polymer of bases is converted into a polymer of protein amino acids, via the protein factories called ribosomes, using the genetic code to translate.

Proteins have remarkably diverse functions, as we've previously noted. Some represent structural components that hold the cell together, similar to the rebar rods that help hold a building together. Many proteins are enzymes; some synthesize various required biochemicals, some do the reverse, breaking down things that are no longer needed. Some of the most interesting proteins are called transcription factors. Their job is to bind to specific sequences of DNA and regulate the expression of other genes. These are the proteins that make sure the right genes are expressed at the right times. Some of the genes regulated by transcription factors are additional transcription factors, which will in turn regulate more genes. It is easy to see how transcription factors can thereby initiate genetic cascades, with hundreds or more genes eventually altered in expression level. One transcription factor gene activates other genes, including more transcription factor genes, which activate still more genes, including more transcription factor genes, and so on.

This interwoven connection of genes, with some regulating others, and some of the others regulating still more, is referred to as a genetic regulatory network. Such genetic programs drive the formation of our different body parts during development. Genetic regulatory networks have a strong similarity to the neural networks of the brain. The cells of the brain, the neurons, are wired together in a complicated circuitry called the neural network. In the nervous system a single neuron receives input from multiple upstream neurons, through its dendrites. It averages these signals, some stimulatory and some inhibitory, and when sufficiently activated it will send an output signal through its axon, which can branch out and influence the activation of multiple downstream neurons. The gene is quite analogous to the neurons, receiving input from many upstream genes. The proteins encoded by these upstream genes bind to the regulatory regions of the DNA. A typical gene will have binding sites for numerous transcription factors, some of which are activators and some repressors. The gene, like the neuron, averages these input regulatory signals, and if there is sufficient net positive input it fires by expressing its own gene-specific protein, which then impacts expression of additional downstream target genes. These networks, both neural and genetic, can be enormously complex. The genetic regulatory network drives appropriate expression levels of tens of thousands of genes, while the neural network sets appropriate activity levels for the approximately one hundred billion neurons of the adult human brain.

The neural network of the brain can accomplish some incredible effects, including memory, intelligence, and self-awareness. The genetic regulatory network is also able to do some quite amazing things, like direct the transformation of a single cell, the fertilized egg, into a complete human being.

DNA is copied into RNA, which drives the synthesis of protein. This is what we are all taught in our biology courses. But now that we have sequenced the genome we know that this isn't the whole story. Indeed, it turns out that of the three billion bases of the human genome, less than 2 percent have this function of encoding protein. That means that about 98 percent of our DNA does something else! So what on earth is the rest of the DNA doing?

INTRONS

The base lettering of the DNA is continuous, from one end of each chromosome to the other, and the amino-acid sequence of proteins is also continuous, from one end to the other. It would seem to make the most sense to have one section of the DNA molecule, a gene, continuously encode the amino-acid sequence of a protein. But, to our considerable surprise, it turned out that this is rarely the case. Instead, the coding sections of DNA are disrupted by noncoding sections, which don't code for amino-acid sequence at all. So, for an individual gene, as you read along the DNA sequence, you will find a stretch of DNA that encodes the first amino acids of the protein, followed by DNA with no coding function. Then, after the noncoding region, which can sometimes be very long, the coding sequence will resume, again dictating some of the amino-acid sequence of the protein. For some genes there will be only one noncoding interruption, and for a small percentage of genes there is in fact no interruption at all. But for most genes the coding sequences are interspersed with many noncoding regions of DNA. These noncoding sections of genes have been termed introns, with the word derived from *inter*rupting cod*ons*. That is, the introns interrupt the codons.

This creates a problem, with the RNA copied from the DNA carrying coding sequences separated by multiple intron sequences. The cell has to get the coding sequence continuous, somehow, so the ribosomes can properly convert it into the amino-acid sequence of proteins. This is accomplished by a process called RNA splicing. A set of splicing proteins can recognize the junctions between coding sequences and introns, specifically remove the intron sequences, and splice together the coding pieces of RNA. The blocks of sequence that end up in the final RNA after splicing are referred to as exons, because they are the *ex*pressed sequences, with the separating intron sequences discarded.

This appears a very convoluted way to achieve gene expression. Why are not all of the coding sequences of the

gene continuous in the DNA, avoiding the necessity of rec-
ognizing and removing noncoding regions? In fact, for most
single-cell microorganisms, like bacteria, this is indeed the
case, as their genes do not have introns, but for higher organisms whose cells have a true nucleus (the eukaryotes),
including people and all other animals, we find that they all
have genes with introns.

One advantage of having introns appears to be increased
flexibility of gene expression. Most genes of the human
genome, for example, can have their copied RNA spliced in
more than one way. Say, for example, that a gene has ten
exons and nine interrupting introns. We used to think that
the RNA splicing process would just splice out the nine introns, leaving the ten coding sequences to be translated into
protein. But for most genes this simple story is not always
the case. Instead there is a remarkable flexibility in the RNA-
splicing process. For example, the RNA-splicing machinery
might sometimes skip over exon 9, and directly splice exon
8 to exon 10. This of course has a dramatic effect on the encoded protein, with the absence of exon 9 changing the encoded protein, deleting some of the internal amino acids.
This alternative RNA processing is not random, but tightly
controlled by mechanisms that are still not fully understood. In some cases the RNA of a gene will be processed
one way in one cell type, and another way in a different cell
type. Most mammalian genes, about 90 percent, according
to the latest estimates, seem to have this plasticity in terms
of processing possibilities.

As mentioned, one of the most unexpected results of the
sequencing of the human genome was the discovery that
people have far fewer genes than we'd thought. Even before
the sequencing we'd known that most of the DNA does not

consist of genes. By measuring the total amount of DNA per cell we knew there were about three billion base pairs of DNA there, which could encode as many as three million genes. The amount of DNA per cell is called the C-value. There is a phenomenon referred to as the C-value paradox, which notes that certain "lower" life forms actually have higher C-values (more DNA) than people. For example, some amphibians, like frogs and salamanders, and some plants, can have five or ten or more times as much DNA per cell as a person. It didn't seem reasonable to think that these lower life forms would require more genes than people, so it appeared likely that some significant percentage of the DNA might not represent genes.

It turns out that some of this extra DNA, which doesn't encode protein, has a regulatory function. These are sequences that are responsible for getting genes turned on and off at the right times. You don't want genes that should just be active in the brain turning on in the liver, for example. And besides simply regulating on or off states, it is important to control the level of expression of genes, so that the throttle, so to speak, is set just right. Every gene has an appropriate expression level setting for every cell type, in every organ, in every stage of development.

We've touched on this regulatory process before. Certain genes encode proteins called transcription factors, which bind to precise DNA sequences in the genome, located next to specific genes. When the protein attaches itself to the binding site, it regulates the expression of the adjacent gene, either turning it on or turning it off.

It should be emphasized that this is a very simplistic view of gene regulation, a process that in reality is incredibly complex. Each gene has many binding sites for many regulatory proteins, so there are typically a large number of inputs, some positive and some negative. In addition, some proteins bind only very weakly to the DNA recognition site, and in order for them to attach firmly they require the presence of another bound protein, which they also bind to. And many proteins actually must first bind to other proteins before attaching to DNA. In these cases the binding sites for the two proteins must be found at just the right distance from each other on the DNA to allow simultaneous, and therefore stronger, attachment. And some proteins don't bind to the DNA at all, but only interact with proteins that have previously bound to the DNA. Complexities abound.

So the regulatory processes that control the expression of even a single gene can be exceedingly intricate. It is not uncommon for an outstanding research laboratory to spend over a decade studying the regulation of one gene. The sequences that flank a gene and promote its expression are called promoters, and this process of studying the regulation of a gene is often referred to in the field as promoter bashing. Studies of this sort have often been extremely informative. A laboratory might start by studying the regulation of a seemingly boring gene, perhaps encoding an enzyme or a structural protein of no striking importance, thereby identifying critical transcription factors that regulate this gene. The study then could shift to understanding the regulatory proteins, finding the other downstream target genes they control, and also finding the genes upstream of the regulators—finding the regulators of the regulators. This sort of investigation eventually defines important genetic regulatory networks, helping us better understand, for example, how genes drive the formation of organs during development.

It has been proposed that gene regulation is actually more important than the genes themselves, when it comes to understanding what sets the various species apart. Maybe it isn't so much the genes that you have, but what you do with them. We know that all mammals have pretty much the same set of twenty-five thousand genes. And we know that about one-third of the proteins encoded by the genes of chimps and people are exactly identical. Some of the species' differences are surely the result of the remaining two-thirds of proteins that are somewhat different in sequence, but a significant percentage of them could be the result of different genetic regulatory networks that cause differences in gene-expression patterns during development and in the adult.

The current DNA-sequencing revolution will resolve this issue, by giving us the complete sequences of thousands, and then millions, of different people. By having sequences for so many individuals we will be able to directly look and see if sequence differences that correspond with specific individual attributes are more often found in coding sequences, which determine proteins, or in regulatory sequences, which control gene-expression patterns. Both will be important, but we can't predict at this time which will turn out to be the more common determinant of attributes such as looks, intelligence, and athletic ability.

So some of the DNA encodes proteins, some of the DNA is found in introns, and another fraction performs a regulatory function. How much of the DNA is regulatory? We aren't sure at this time, as coding sequences are far easier to identify than regulatory elements in the human genome sequence. But based on the relatively small number of genes that have been extensively "promoter bashed" we can extrapolate that the amount of DNA devoted to regulation is not likely significantly greater than that devoted to coding. If 2 percent of our DNA encodes proteins, and perhaps 3 percent is regulatory, this means that we still have about 95 percent of the human genome that is not coding or regulatory, but does something else.

7

JUMPING GENES
(TRANSPOSABLE ELEMENTS)

The human DNA sequence provides a remarkable landscape to the molecular geneticist. In one sense it is like a large desert, with a scattered oasis here and there, representing the relatively few coding sequences of genes. But the desert is not entirely empty. It is not a vacuum, like outer space, but instead is occupied by a variety of sequences that are quite interesting in their own right. Among the most peculiar are the transposable elements, or "jumping genes." Unexpectedly, it turns out that transposable elements make up a much larger percentage of the genome than traditional genes.

Barbara McClintock first discovered these strange genetic elements, before Watson and Crick had even defined the structure of DNA. She did classical genetics experiments, mating different types of corn. She found that for certain matings the ears of corn that resulted would have curious coloration patterns—not only would different kernels have different colors, but even different cells within a single kernel would have different colors. Through a series of brilliant experiments, which would win her the Nobel Prize, she demonstrated that discrete little pieces of the chromosome were capable of moving around from one chromosomal position to another. She called these mobile regions controlling elements, because they could, under certain circumstances, regulate expression patterns of other genes, including those that synthesized the pigments giving corn its color.

In modern jargon these controlling elements are referred to as transposable elements because they can move, or transpose themselves, from one place to another in the genome. This is difficult to comprehend. How can an integral part of the DNA sequence, usually from a few hundred to a few thousand bases in length, shift from one place in the chromosome to another? What allows it to jump about?

The answer is that there are proteins, in this case enzymes, that are specifically designed to recognize the sequences of the transposable elements and to facilitate their movement. These enzymes allow the transposable elements to make extra copies of themselves, which can then float around in the nucleus, and insert themselves willy-nilly, with no apparent pattern, into new positions in the DNA sequence. And where do these special enzymes come from? These enzymes that allow transposable elements to move? The transposable elements themselves encode these proteins. Indeed, that seems to be their only function, to encode proteins that drive their own movement as well as their replication.

Most of these movements occur on an evolutionary

timescale. The rapid jumping events seen by Barbara McClintock following certain matings in corn are actually very rare. More often they move very slowly, maybe a little bit each generation. Nevertheless, the cumulative effects over evolutionary time can be impressive. In the human genome, over 30 percent of all the DNA sequence consists of transposable elements. And remember that only 5 percent of the sequence is devoted to genes and their regulation. So, at least six times as much DNA sequence encodes transposable elements.

Do these transposable elements do any good? Are they harmful or helpful? This is still a matter for conjecture. It has been proposed that they might have an evolutionary function. For example, sometimes when they move they can carry blocks of chromosomal DNA around with them. This gives the genome a more plastic character, generating new genetic diversity for natural selection to act upon. It is definitely true that some transposable elements are able to shuffle exons about the genome, producing genes with novel exon combinations, which in some cases might be of particular use. There is also evidence that certain transposable elements can respond to very stressful conditions, activating and accelerating such genome modifications that might provide new, more fit gene combinations. But, as for all natural mutation events, which are random and not directed, the vast majority are harmful and not beneficial.

These transposable elements can be viewed as submolecular parasites. It appears that there is evolution going on within the microscopic world of DNA. Transposable elements can preferentially replicate, making more copies of themselves, making themselves more abundant in the DNA sequence. This is akin to the species in the wild that is particularly adept at reproducing and populating its environment.

Consider a single transposable element. It can make multiple copies of itself, which move about to new random positions in the DNA sequence. And each of these offspring is then capable of making additional copies, which occupy still more positions in the DNA. And so on and so on. It would seem that if some DNA sequences could successfully out-replicate others, then eventually they should make up essentially the entire DNA sequence. Perhaps it is surprising, when viewed from this angle, that even more of the human DNA sequence isn't made up of transposable elements.

But of course, as with any parasite-host relationship, there must be a balance. If the parasite is too overwhelming then it can kill the host, and in the process kill itself. So, while the host is trying to control the parasite, the parasite itself is also trying to exercise self-control, to avoid destruction of the host. As the number of copies in a family of transposable elements goes up there is often a repression mechanism that kicks in to reduce the rate of expansion. In addition, the host cells have mechanisms, still very poorly understood, for the inactivation of repeat sequences like transposable elements.

Why are transposable elements important? First, if we want to be able to relate DNA sequences to traits, then it is essential to understand the functions of the different types of DNA elements. If a sequence difference between two people shows up in the middle of a transposable element, for example, then it is unlikely to be related to their trait differences, such as their different weights, but we must be able to recognize the transposable elements to know this.

Second, transposable elements have some very interesting evolution stories of their own to tell.

THE INVASION OF THE BODY SNATCHERS

Scientists who work with fruit flies have uncovered a very extraordinary transposable element event, a family of jumping genes that suddenly appeared out of nowhere, peppered the fruit fly genome with multiple copies of itself, scattered about randomly in the DNA sequence, and then repressed itself, to prevent further damage to the host. It provides an elegant example of how these little DNA parasites can infect a species, wreaking havoc, and then come under control, resting in their newly occupied DNA homes. It shows in dramatic fashion the competition between the host and the DNA parasite for survival.

When fruit-fly geneticists started their work, in the early 1900s, one of the first things they did was to establish isolated laboratory stocks of flies. These lab populations were started from wild flies from many locations around the world, including Oregon and Hawaii. Despite their diverse geographic origins they were all the same species of fruit fly, called *Drosophila melanogaster*. The geneticists would grow these flies in the laboratory, mutate them with chemicals and X rays, mate the flies to each other to study inheritance patterns and to observe the effects of the mutations, and work out the principles of genetics, which apply not only to fruit flies but also to people.

In the 1970s, with the advent of recombinant-DNA technology, DNA cloning, and primitive sequencing, it became possible to expand these studies to include analysis of the composition of the DNA sequences present in the genome. We now had some basic tools that allowed us to study families of sequences. We found that some sequences appeared over and over in the genome, not as identical copies, but as similar or related sequences. We could not sequence entire genomes yet, but we could clone individual chunks of DNA using restriction enzymes (special DNA-cutting proteins) and then use hybridization techniques to see how many copies were present, and to find additional DNA clones with similar sequences. If we denatured the DNA, separating the two strands, then different related sequences could come back together with one another, or hybridize, through Watson-Crick base pairing of the complementary sequences. We could radioactively label a cloned copy of one sequence and then hybridize it to other clones, to see how many it would anneal with, because of this sequence similarity.

Occasionally scientists would go back into the wild, to examine fruit flies buzzing around some rotten fruit and compare their DNA to that of the flies that had been maintained in the laboratory. A very surprising finding was made. It turned out that all of the fruit flies of the species *Drosophila melanogaster* in the wild, and not maintained in a protected laboratory environment, had acquired multiple copies of a transposable element family called P-elements. This had occurred over a very short period of time in evolutionary terms, just a few decades. And the phenomenon was incredibly widespread. It didn't matter if the wild fruit flies now came from Oregon, Ohio, Canada, Mexico, Hawaii, Japan, or Siberia.

Where did these P-elements come from? It is an example of what is called horizontal gene transfer. These DNA elements did not come from the parents or ancestors of the flies. They weren't there before. Receiving genes from your parents is referred to as vertical transmission—moving from one generation to the next—and of course this is the standard, as almost all of your genes come from your ancestors. But occasionally something else happens, with genes somehow crossing species' boundaries, moving horizontally from one species to another. But how can a gene jump from one species to another—for example, from a cow to a horse, or from one insect to another? One mechanism is called transduction. It exploits the fact that many viruses are capable of infecting more than one species. Consider the flu virus, with new strains that infect people often coming from animals, such as pigs or chickens. So, a virus can infect a cell, and as it is busy killing that cell and making more virus, a transposable element from the host genome can abandon ship, jumping to the DNA of the virus genome, and hitching a ride. Now, when that virus infects another individual, perhaps of another species, it can jump again, this time to the DNA of the new host. But won't the new cell be killed by the virus, making the trip for naught? This happens sometimes, but in other cases the transposable element's insertion into the virus DNA can functionally inactivate it, making it impossible for it to further replicate, allowing it to infect a cell, inserting the viral DNA, but not allowing replication and resulting lysis (disintegration) of the new cell. In this case the virus has effectively transmitted genetic information from one species to another. Transduction can, with low efficiency, move any piece of DNA from one species to another. But for transposable elements, with their innate ability to move onto and off of the viral genome, the efficiency is dramatically higher than for your average gene.

One result of this transduction is an interesting exchange of genetic information across species boundaries. There is a genetic cross talk, albeit rare, that moves DNA sequences from one species to another, creating new genetic diversity for natural selection to act upon. This process allows the various species of the planet to share their DNA with one another.

Such horizontal transfer of DNA can result in surprising distributions of sequences across different species. We typically expect to see that closely related species share similar DNA sequences and closely related genes. If two species recently diverged from a single common ancestor, then their DNA should still be very much alike. But sometimes, for some genes, especially transposable elements, a quite spotty distribution is observed, with, for example, *Drosophila melanogaster* having P-elements and another very closely related species not having any. This is, of course, because the common ancestor of the two species did not carry the P-elements, which were acquired horizontally in very recent time in evolutionary terms, and not all fruit fly species acquired the new sequences.

The P-element story provides a snapshot view of how a genome naturally evolves, for the decades it took for *Drosophila melanogaster* to acquire them is just a snapshot in evolutionary time. Multiply that century by hundreds of centuries, and it is easy to see why the fruit fly DNA is more than 30 percent transposable element sequences.

The story also has a more sinister side, for when the P-elements invaded the fruit-fly DNA, there were significant harmful effects. When P-elements move about, making more copies of themselves and inserting them into new places in the DNA, they create mutations that are generally harmful, inactivating useful genes by disrupting them. When a wild fly with P-elements is mated to a lab fly without P-elements, the elements can sometimes undergo a surge of activity, occupying random positions on the chromosomes from the lab fly. This phenomenon is called hybrid dysgenesis and results in a reduced fitness. That is, the flies are kind of sick.

There was a very puzzling sexual asymmetry to hybrid dysgenesis. When male laboratory flies, without P-elements, were mated to female flies from the wild, with P-elements, nothing happened. But when the reverse mating was performed, using male flies with P-elements and female flies without them, the P-elements underwent rapid movement, causing hybrid dysgenesis. For most matings, when we are studying other types of genes, it doesn't matter which genes come from which parent. The embryo receives the same genes, whether they come from the father or the mother, so it typically makes no difference. But for P-elements it made a huge difference. Why?

It turns out that in the flies from the wild the P-elements have been around for a while now and have come under repression, so they don't move much anymore. These flies make a repressor that blocks the movement of P-elements. When a wild fly makes an egg, the cell is big, with lots of cytoplasm, which includes a fair amount of this P-element repressor. So when a wild-fly egg containing P-elements but also with the repressor joins with a lab fly's sperm, with no P-elements, not much happens, because the repressor blocks transposition events. But in the reverse, sex mating the wild fly with P-elements makes sperm with very little cytoplasm and hence almost no repressor; the sperm then fuses with the laboratory fly egg, which has lots of cytoplasm but no P-elements and therefore no P-element repressor. The result is a cell with unrepressed P-elements that become very active, again moving about and occupying new positions in the DNA.

It seems paradoxical that P-elements, which reduce the fitness of flies when they are moving about and mutating genes, can swiftly spread though a population, quickly inhabiting the genome of every fruit fly in the wild. After all, a foundation of Darwin's theory of evolution is the survival of the fittest. Individuals that are more fit, and have more offspring, pass more genes to the next generation, while genes from less fit individuals should become more scarce with time. Why would P-elements spread when they make individuals less fit?

The answer is multifaceted. The P-elements can be seen as a sexually transmitted disease, inducing rapid P-element mobility and transposition to previously unoccupied chromosomes in an egg with no P-elements. In a way, the P-elements resemble the alien seeds in the movie *Invasion of the Body Snatchers*. The aliens could convert people into aliens while they slept. Then these new aliens could help convert more people into aliens. The end result was, in fairly short order, a world where all the people had been replaced by aliens. The same seems to be true for P-elements. Whenever a male fly with P-elements mates, the chromosomes of all offspring carry P-elements. Because of their ability to preferentially replicate and make more copies of themselves, the P-elements can spread rapidly. We've seen the experiment done in nature.

A mathematical analysis shows that there is a balance between the harmful effects and the preferential replication. Obviously, a fully lethal transposable element would not get far, since it would kill its host and thereby commit suicide. And if a transposable element could not preferentially replicate at all then it wouldn't really be a transposable element, but rather just another piece of DNA, and would only spread through a population if it had some beneficial effect. In between these two extremes exists a wide range where transposable elements can indeed become more prevalent in a population, despite some harmful effects, because of their ability to outreplicate their competing DNA sequences in the genome, and to pass the survival-of-the-fittest test within the microcosm of the DNA genome.

EVOLUTION AND HUMAN DNA CONSTRUCTION

The human genome does not appear to be particularly intelligently designed. Of its three billion bases of DNA, only a small fraction, less than 2 percent, actually encodes proteins. Another small percentage of the DNA sequence is regulatory in nature, responsible for making sure that genes are turned on and off at the right times and places. Another approximately 30 percent is parasitic transposable-element DNA. And this still leaves over 50 percent of our DNA with no known purpose, useful or parasitic. Perhaps this extra DNA provides a protection buffer against the harmful effects of transposable elements. If 100 percent of the DNA encoded essential proteins then every time a transposable element occupied a new genomic position it would kill a gene. But if 98 percent of DNA is noncoding, then 98 percent of the time when a new insertion event occurs it will be relatively harmless. And, of course, it remains possible that some part of this DNA sequence has another function that we just don't understand yet. But it appears quite certain that the mammalian genome, human included, is anything but a streamlined engine of efficiency.

Evidence of evolution abounds in the human genome. We see the genome littered with transposable elements, evidence of historic movements of DNA, with most of these elements now inactive, having accumulated base mutations over time that have rendered their transposase genes nonfunctional. This is like a junkyard filled with decaying cars.

The DNA also carries genomic scars resulting from ancient viral infections. Some viruses, called retroviruses, actually have genes that are made of RNA instead of the normal DNA. But after they infect a cell they use a special enzyme, called reverse transcriptase, which they encode to make a DNA copy of their genes. Howard Temin and David Baltimore shared a Nobel Prize with Renato Dulbecco in 1975 for discovering this phenomenon. It shocked the world of biology to find that RNA could be converted into DNA. It had previously been thought that genetic information only traveled in the other direction, from DNA to RNA. The viral DNA copy is then inserted randomly into the DNA of the infected cell, and it remains there, accumulating mutations with time. Similar retroviral scarring patterns in the DNA of related species provide further proof of their common ancestry. By this measure even the mouse and human genomes are related.

Some types of transposable elements are very closely related to retroviruses. They also encode reverse transcriptase, allowing their RNA transcripts to be copied back into DNA, which then inserts itself into new positions in the cell's DNA. They are like viruses that have lost the ability to make virus particles that spread to other cells and infect them. Instead they just make copies of their DNA, which then "infect" new sites in the DNA of the same cell.

We also see scattered about in the genome what are called pseudogenes. These are copies of normal genes that have also accumulated over evolutionary time multiple random mutations that have inactivated them, so that they no longer code for functional protein. Like most transposable elements and retroviral remnants in the genome, they are evolutionary relics, DNA fossils that speak of the history of our species.

8

GENETIC DISEASE

Woody Allen once said that life is divided into the horrible and the miserable. The "horrible" part of this statement certainly seems apt when one thinks of genetic disease. Consider a person diagnosed with Huntington's disease. Because of the inheritance of a single mutant gene, certain regions of the brain begin to die, usually in middle age, resulting in jerky involuntary movements and mental decline. The severity can vary, depending on the form of the gene inherited. Early symptoms include dramatic mood swings, depression, irritability, trouble learning new things, memory lapses, and difficulty making decisions. As the disease inexorably progresses, the person has increasing difficulty concentrating on intellectual tasks, and eventually has difficulty eating and swallowing. There is a slow progression involving ultimate dementia and death. This horrible disease has touched me personally, as I once lived across the street from a house where the first thing you noticed when walking in was a foul fecal odor, and where there was a woman who could no longer function as a wife and mother. And today a co-worker, still young, brilliant, beautiful, and vibrant, and known to carry an altered form of the Huntington gene, is beginning to show early signs of the disease.

Huntington's disease is just one example of over a hundred diseases for which we know that the mutation of a gene is the underlying cause. A few additional examples include cystic fibrosis, ALS (Lou Gehrig's disease), hemophilia, Marfan syndrome, sickle-cell disease, SCID (bubble-boy disease), neurofibromatosis, Tay-Sachs, and Rett syndrome. Each may seem a simple innocent name of a disease, but each can be a personal disaster multiplied by the thousands, or millions, of people individually affected.

Consider Tom Cramer, a fictional composite example of a teenage boy with Lesch-Nyhan syndrome. This disease is caused by mutation of the gene encoding an enzyme abbreviated HPRT (which stands for hypoxanthine-guanine phosphoribosyltransferase). The loss of both functional copies of this gene has devastating consequences, including severe behavioral abnormalities. In particular there are self-mutilation behaviors, with head banging, finger biting, and lip biting. It is not unusual for people with Lesch-Nyhan syndrome to actually bite off their own lips and fingers. This destructive behavior can also be turned on others, with kicking, head butting, and spitting. Tom therefore requires many protective restraints, including a helmet. He uses a wheelchair with no exposed hard metal surfaces, and his arms are kept secured to the sides of the wheelchair. Indeed, he even sleeps in a special protective bed. Interestingly, these patients do not want to hurt themselves and are fearful when left without supervision or protective restraints. And when they hurt others they are generally extremely apologetic, although they will then proceed to do it again. There is clearly an internal battle going on that they cannot win. Tom requires continual supervision. Yet despite these problems, and moderate mental retardation, Tom is cheerful, interactive, fun loving, able to attend high school, and is loved very much by his family and friends.

Every type of genetic disease (and again, there are over a hundred currently known and likely many hundreds more yet to be identified) tells a dismal story, some more heartbreaking than others. The goal here is not to detail them all, but rather to appreciate the astounding significance of our coming ability to eliminate them. In the not too distant future it will be possible to completely sequence the DNA of every living person, and to thereby define every gene mutation carried. The truth is, we all have mutations, but usually in only one copy of a gene, and for most genes it is perfectly fine to have one good copy and one bad copy. The problem arises when two people with a mutation in the same gene marry. Then there is a 25 percent chance that a child will get two bad, mutant versions of the gene, one from each parent, and as a consequence suffer genetic disease. But if we know our complete DNA sequences then we can be on guard for this eventuality, perhaps by restricting who we marry, or perhaps more likely by screening embryos through DNA sequencing when the danger of severe genetic disease is present. And in time, as such genetic screens become more common, it might be possible to completely remove such harmful versions of genes from the human population.

NATURE VERSUS NURTURE

While we can appreciate that at some point it will be possible to choose the genes of our children, perhaps we nevertheless still question the importance of those genes in defining their characteristics. Let us consider the often debated issue of nature versus nurture. What determines the attributes of a person, their genes or their environment? While it is clear that bad genes can cause genetic disease, aren't the rest of us, with normal genes, more shaped by our surroundings, our friends, our schooling, and our socioeconomic status?

Of course both genes and environment are important. To determine their relative contributions scientists have turned to the study of twins. We know that identical twins have the exact same set of genes, since they are derived from the splitting of a single early embryo. Fraternal, or nonidentical twins, on the other hand, only share 50 percent of their genes, the same as other brothers and sisters. Yet both kinds of twins, identical and fraternal, are usually raised together, are the same age, have the same parents, and live in very similar environments. So one can ask, how much more similar to each other are identical twins, compared to fraternal twins? This begins to define the genetic contribution.

Another way to address this question is to study twins that were for some reason separated at birth, and therefore raised in different environments. Are identical twins, with identical genes, still very similar to each other, even when raised in different environments? Are they more similar to each other than fraternal twins raised in different environments? If genes really were important in defining characteristics, then one would predict that identical twins raised in different families would still show significant similarities, above and beyond their physical appearances.

Sir Francis Galton was the first to use twins to study the genetic contribution to intelligence. He was a cousin of Charles Darwin's, and was very much taken by the theory of evolution and interested in studying the inheritance of characteristics, including intelligence. He coined the phrase "nature versus nurture" and concluded from his work that nature was more important than nurture.

The modern-day Minnesota Twins Study is one of the most comprehensive studies of this sort. It is an analysis of hundreds of identical (monozygotic) and fraternal (dizygotic) twins raised in the same or in different environments. Of course, as would be expected, there is a very strong genetic component to physical attributes, such as height, weight, and appearance. If you take identical twins and rear them apart, they still look identical. And if you take fraternal twins and raise them together, they still don't look the same, although there will be similarities, in part because they do share half of their genes. In addition to appearance, IQ also showed a very strong heritability, with the results indicating that about 70 to 80 percent of a person's IQ is determined by genes and only 20 to 30 percent by environment.

There were also some surprising results to come from the study, showing a genetic contribution to unexpected psychological attributes. For example, identical twins, even when reared apart, scored more alike in measures of happiness than fraternal twins. They seem to share a genetic constitution that to some degree preset their level of happiness. And another surprise was the connection between genes and level of religiosity. One might expect that the family environment in this case would be particularly important, with children exposed to religion on a weekly basis much more likely to become more religious. But if one identical twin was more religious than average, then the other was likely to be as well, even when raised separately. Of course the genes did not pick the faith, just the level of involvement in religion. One twin might be a devout Protestant, while the other twin, raised by a different family, would be a devout Jew.

There is also a genetic component to other psychological features, including mannerisms and job preferences. For example, the first two identical twins enrolled in the Minnesota Twins Study were James Arthur Springer and James Edward Lewis, who had been adopted out at the age of one month, and not reunited until the age of thirty-nine. It turned out that each had married and then divorced a woman named Linda, and then remarried a woman named Betty. They shared common interests and hobbies, including mechanical drawing and carpentry, and in school they had the same favorite and the same least favorite subjects. Further, they both suffered headaches, at the same time of the day, and they both smoked and drank the same amount. Other twins also showed striking similarities, including Oskar Stohr and Jack Yufe, who were separated at six months of age, with Oskar brought up as a Catholic in Germany, and a member of the Hitler Youth, while Jack was raised a Jew and lived for a while in Israel. These twins, reunited in their forties, showed similar tastes in food, similar speech and thought patterns, similar gaits, and shared idiosyncrasies, including always flushing the toilet before they used it.

In summary, genes play an important role in defining

both our physique and our psyche. But it is also important to note that they don't completely define us. Identical twins are certainly similar, yet despite having the same genes they are not the same people, with all the same attributes. Although our genes establish limits, within those limits there exists a wide range of possibilities, determined by chance and our environment. What we are, of course, in the end is the result of both nature and nurture. But good genes are an obvious advantage.

9

THE SURPRISING EMBRYO

The making of a human being is quite an amazing process. It all starts with the formation of the eggs and sperm. The scientists who study these cells argue that they are clearly the most important cells of the body. The eggs and sperm are the vehicles that carry DNA from one generation to the next. Without them the species stops, period, and is no more.

Each egg and sperm has the potential to make a person. They just need to join with an egg or sperm of the opposite sex. And from a genetic perspective each is unique. Every other mammalian cell has two copies of each gene, but eggs and sperm have only a single copy of each gene, so that when they fuse together the normal two-copy state is restored. The genes are located on the chromosomes, each of which is primarily a very long DNA molecule. People have twenty-three pairs of chromosomes. We each inherit one set of twenty-three chromosomes from our father, and another set from our mother. During meiosis, the process of making eggs and sperm, there is a random distribution of chromosomes. You have two copies of every chromosome but only one copy of each goes into an egg or a sperm. And which copy of each chromosome pair moves to an egg or a sperm is unrelated to which copy of another chromosome pair joins it. Since we have twenty-three chromosome pairs this creates an enormous number of possibilities.

The process of recombination creates even greater genetic variety in eggs and sperm. Special types of cell divisions are necessary to create eggs and sperm, since they have half the normal number of chromosomes. At one stage in this process the two copies of each chromosome pair line up next to each other. The chromosomes are named with numbers. So the two copies of chromosome number 1 will come together, and so will the two copies of chromosome number 2, and so on. While they are next to each other the chromosomes undergo a process of recombination, randomly shuffling their genes by exchanging regions of DNA. The net result is that each egg and sperm carries a different combination of genes from all the others. This gene-swapping capacity can be harnessed by genetic engineers to genetically modify cells, to be discussed in a later chapter.

The percentage of eggs and sperm that are wasted is incredibly high. Females are born with perhaps a million primordial eggs, of which about five hundred will actually ripen and be made available for fertilization, and of these only two, on average in a stable population, will actually be fertilized and contribute to the subsequent generation. Each couple needs to only produce two children to replace themselves. For males the numbers of sperm wasted are even more staggering. It is estimated that a male will produce about ten trillion sperm in his lifetime, of which only about two, on average, will be used.

When you think about it, your chances of existing seem absolutely infinitesimal. Of course people keep reproducing and making more of themselves, so it is not surprising that the planet is full of people. But what is the likelihood that you would be among them? Even given your two parents, what are the odds that the one specific egg from your mother, out of hundreds possible, and that one precise sperm from your father, out of trillions possible, would come together to make you? Obviously if a different sperm had won that race to the egg, then a different person would have resulted, and not you. When your parents were born, what were the chances they would meet each other and marry? And for your parents to even get born they had to beat the same odds you did, and have the exact right egg get fertilized by the one correct sperm, from their parents. And then consider the parents of your parents, and their parents, and so on, for a hundred generations. In each case the right two people had to meet and mate, and the right egg and sperm, out of many trillions of possible combinations, had to come together. One break in that incredibly improbable chain of events, and you would not exist.

THE EARLY EMBRYO

When that specific egg and sperm do come together they fuse to form a single cell—the zygote. Of course we all began as a zygote. Everyone does. This fertilized egg is very small, just barely visible to the naked eye. Even a whale starts as a tiny zygote, about the same size as for other mammals. Under the microscope a zygote is rather unremarkable. Like other cells it has a nucleus, where the DNA resides, and a surrounding cytoplasm, where proteins are made. Its most unusual feature is that it is very large compared to most cells, even though it is just a speck, tinier than the period at the end of this sentence. But the fertilized egg doesn't at all resemble a person. There is no face, no head, no brain, and no heart. This fact was not always appreciated. Early embryologists mistakenly thought the embryo might begin as a tiny person, with arms, legs, and other typical features. And then this tiny person would just grow bigger.

What the human zygote does have that sets it apart from all other cells is a remarkable potential. It can turn itself into a person. And this is indeed an astounding transformation from a microscopic spot of protoplasm, a single cell, to a very large mass, generally well over a hundred pounds, with trillions of cells, and hundreds of different cell types, all interacting together as a team.

This process of converting a single cell into a person is referred to as development. It is driven by a genetic program encoded in the DNA in the nucleus of the zygote. It is mind-boggling to consider the power of this genetic program. There are only about 25,000 genes in the zygote. How can these few genes direct the formation of a person's many complex organs, including the enormously intricate brain? We certainly don't have all of the answers to this question, and thousands of research scientists around the world are devoted to better understanding this fascinating process.

The creation of babies with desired gene combinations requires the laboratory manipulation of early human embryos, up to a few days old. The degree of manipulation would depend on the level of selection and even gene modification desired. At the low end we could do what is already done today in many clinics around the world, only armed with some additional knowledge of the relationship between DNA sequence and physical traits, and therefore greater power to select for more than just the absence of disease. Following in vitro fertilization the embryos would be grown in special media, or food, for a few days in an incubator, keeping them at body temperature, until they were at the eight-cell stage. Then a single cell would be removed from each embryo, their DNA analyzed, and the embryo with the optimal gene combination selected for insertion into the uterus of the mother, where it would develop into a baby. This is similar to what was done by the Nash family to save their daughter Molly, but a better understanding would allow one to select on the basis of multiple traits, and not just for the absence of disease genes and transplant compatibility. At the high end of the manipulation-and-selection scale we could take early embryos—again, following in vitro fertilization and growth for a few days in the laboratory—and use stem-cell and gene-modification technologies, described in the following chapters, to more actively create the desired gene combinations.

To begin to better understand how all of this works, let's briefly review early human development. As we've noted, the sperm fuses with the egg to accomplish fertilization and to produce the zygote, which carries equal genetic contributions from each parent. Perhaps surprisingly, the early mammalian embryo is surrounded by a shell, called the zona pellucida. So at the day of fertilization, called day zero, we see a single cell, with a small nucleus in the middle, and with an outer shell, as shown in the diagram on the next page. Early mammalian development is very slow. It takes about a day for the fertilized egg to divide into two cells, of course each with its own nucleus, and still surrounded by a shell. And then another day for these two cells to divide again, into four cells, and another day to reach the eight-cell stage. These first three days, and first three divisions, are what we are most concerned with, as this is the time frame involved for most of the required designer-genes procedures. However, for a few more advanced schemes it is necessary to let the embryo go a couple more days and a few more cell divisions, to day five, to make a blastocyst, a hollow sphere with a clump of cells attached to the inner surface. This clump of cells, called the inner cell mass, is what goes on to make the baby, and it can also be used to make stem cells, which are capable of turning into any cell type in the body. The inner-cell mass cells are relatively small, with very little cytoplasm. Their nuclei are shown as black ovals in the day-five embryo in the diagram.

Early mammalian development, including human, is highly regulative. In simple terms this means that the process is surprisingly smart and flexible. During the first three days of development, up to the eight-cell stage, if any cell was somehow separated from the rest it would have no problem moving ahead alone and making a baby. Indeed, this is one way to make identical twins, triplets, etc. The cells at these early stages are not firmly attached to one another, but more like a cluster of grapes, barely touching. So it is quite possible, although still rare, for cells to separate. Another way to make identical twins is to somehow separate the single clump of cells making up the inner cell mass of the blastocyst into two clumps. And of course since all of

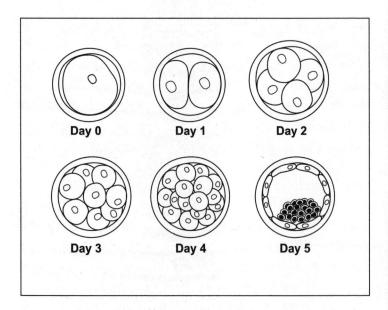

Day 0   Day 1   Day 2

Day 3   Day 4   Day 5

the cells of both resulting embryos are derived from the single original zygote, they are all genetically identical.

In the laboratory such embryo manipulations are quite straightforward. It is possible to take an eight-cell embryo and to separate the cells in any manner desired. This could mean making eight individual cells, or four pairs of cells, or any other possible combination, such as five cells plus three cells. Then each individual cell or group of cells can proceed to make a perfectly normal person. It is therefore possible to produce from a single eight-cell embryo as many as eight people, as few as one, or any number in between.

The converse can also happen, although it is more rare. Occasionally the mother will release two eggs at once, which if fertilized would normally give rise to nonidentical, or fraternal, twins. But sometimes the resulting two early embryos, up to the eight-cell stage, join together. The result is a single person that is actually a mixture of two people, referred to as a chimera. This term comes from Greek mythology, in which a chimera was an animal with a lion's head, a goat's body, and a serpent's tail.

The concept of a chimera is more difficult to grasp than that of identical twins, which we've all experienced. A chimera is really two people in one. For example, consider two eight-cell embryos being manipulated in the laboratory. At one extreme, if each individual cell is separated from the others, then sixteen people result. But if the two early embryos are combined together, forming one ball of cells instead of two, the result is one embryo that goes on to make a single person, a chimera. So two eight-cell embryos are capable of making as many as sixteen people, or as few as one.

Human adult chimeras usually just look like average people and it takes some very careful analysis to detect them. They often show up by accident during parentage testing, when the DNA analysis seems to indicate that a mother did not bear her biological son or daughter. This is because the chimeric mother is a mix of two genetic types of cells. One DNA type of cell contributed the egg that gave rise to the offspring, but the other DNA type of cell was used for the DNA testing, so the true parent and the offspring did not match.

Mouse chimeras are useful research tools for developmental biologists, who will often mix the cells of a mutant embryo, which on its own would be too genetically defective to go very far in development, with cells of a normal embryo. The researcher will then study the chimeric embryos, and determine which cell types the mutant cells are capable of making, when helped along by the normal cells.

In the laboratory it is even possible to make chimeras

from more than two independent embryos. Indeed, one can take five early mouse embryos, perhaps from five different mothers, and push them together under a microscope, and they will join into one conglomerate of cells and make a single chimeric mouse. Typically this mouse would appear perfectly normal, not physically different from any other mouse. You wouldn't know that it had ten parents, instead of the usual two. But if one joins embryos that all have distinct genes for coat color, then the cells from the various contributing embryos go on to make different random regions of the skin, for example giving some areas that are black, some that are brown, some white, and so on, reflecting their diverse parentage. The result is a rather remarkable chimera, whose coat is a splotchy jumble of many different colors.

The results described above are not what one might intuitively predict. For example, one might expect that if you had a two-cell embryo, and separated the two cells, each cell might now make only one half of an embryo. One cell might, for example, make the head and chest, and the other cell the stomach and legs. And this sort of thing is indeed seen for embryos of some simpler organisms. But for mammals the embryonic cells seem somehow more intelligent. It is as if they can look about and see what other cells are there and act accordingly. If the individual cells of an eight-cell mammalian embryo are somehow separated, then each can see that it is now alone, and must do the entire job of making a complete organism, be it mouse or man. On the other hand, if five embryos are packed together, then the cells somehow see that the team is now big, and all of the cells must work together to make but a single individual, and not five. It should be noted that this plasticity works for only early embryos that are just a ball of cells. If you take older, more developed embryos and push them together they just continue to make multiple individuals.

These properties of early embryos raise a number of interesting ethical questions. For example, when we take a human eight-cell embryo, and remove one cell for DNA diagnostic procedures, which is done routinely today in many clinics around the world, are we committing murder? Of course the remaining seven cells can go on to make a perfectly normal person, with no damage resulting from the removal of a single cell. But the single cell that we tested was in fact also capable of making a person, a twin of the person resulting from the seven-cell embryo. So we in essence sacrificed one embryo, the single cell, to genetically define the remaining embryo of seven cells. Ethical considerations of this sort will be discussed more in a later chapter.

10

STEM CELLS

Stem cells are almost magical in their power. Imagine a single cell that has the ability to turn into almost anything—to become a heart cell, or a pancreatic insulin-producing cell, or a brain cell. Imagine the good that could be accomplished with these cells, by repairing a heart damaged by a heart attack, or repairing a brain damaged by stroke, or curing diabetes. The possibilities for regeneration and repair of diseased or damaged organs are almost endless. Stem cells could truly usher in a new age of medicine.

Where do stem cells come from? Perhaps not surprisingly, some of the most powerful come from the early embryo. After all, the fertilized egg is the most potent stem cell of all, able to give rise to all of the cell types of the body, as it proceeds to make a complete person. If we let the fertilized egg grow just a few days in culture, it will then form the early embryo called the blastocyst. As mentioned previously, this embryo, resembling a basketball in shape, has an outer sphere of cells, an inner empty space, and on the inner surface, adhering to the sphere, a clump of cells rather unimaginatively called the inner cell mass. The outer sphere of cells ends up making supportive structures, like the placenta, while the inner cell mass eventually becomes the person.

It turns out that we know how to take these inner-cell-mass cells, place them on plastic petri dishes, feed them the right food, set them in an incubator at the correct temperature, and grow them in the laboratory. Indeed, under the right conditions they stay inner-cell-mass cells, and don't proceed to differentiate, but do continue to divide, providing a continuous supply of cells. They are perpetual early embryo cells, and are referred to as embryonic stem cells. Other, generally less powerful, stem cells can be found in more advanced embryos, or even in adults.

So, what good are stem cells? At present there are a limited number of effective stem-cell therapies available for people. The most common, and most proven, is the blood-stem-cell transplant, as discussed earlier in the case of Molly Nash. This can be accomplished through a bone-marrow transplant or by using umbilical-cord cells, which also have blood stem cells. This sort of transplant can restore the ability of a person to make both red blood cells, which carry oxygen, and the blood cells of the immune system, needed to fight infections. Stem-cell therapies have also been used in people to treat diabetes, heart disease, and kidney disease, with promising results. Another use of stem cells is to correct corneas of the eyes that have been scarred by infection or trauma. But we are really in just the beginning phases of learning how to use stem cells to treat human disease. In mouse models researchers have used stem cells to treat spinal injuries, Parkinson's, and other devastating conditions, with encouraging outcomes. Over the past few years legions of scientists have worked out conditions for converting mouse embryonic stem cells into many different types of cells that could be used to help patients. We need to further extend this work to humans. In the future these cells could provide a powerful elixir for the treatment of many diseases, helping us extend useful life spans dramatically, and, for example, perhaps even allowing paraplegics to walk again.

So what is the downside? Why the controversy? Well, these stem cells are often made from early embryos, meaning embryos must be sacrificed to produce them. Blood stem cells, which can be recovered from adult bone marrow or newborn umbilical cords, are the exception. So to make human stem cells that could be used for the treatment of most conditions it is generally thought that we need to sacrifice human embryos, a matter of considerable ethical debate. In the midst of this debate it is often overlooked that embryos at this same stage, blastocysts, are routinely sacrificed by the tens of thousands in in vitro fertilization clinics around the world, with relatively little fanfare. But for some reason the notion of doing the same thing for the sake of stem-cell research has sparked considerable anger.

How can we move past the controversy and on to the rewards of stem-cell research? There are several options. Perhaps the least likely course of events is to finally settle the argument concerning human embryo sacrifice and to push ahead. It is true that many embryos at this same early stage are naturally lost during normal reproduction. A surprisingly high percentage, about half, normally fail to implant (lodge in the wall of the uterus). Without implantation the embryos cannot form a placenta; they therefore cannot feed, grow, and develop, so they starve and perish. This is another topic we will return to in a later chapter. It has also been argued that we could just use the leftover blastocysts from in vitro fertilization clinics, instead of throwing them away, to make additional human stem cells.

But of course these embryos are not just worthless balls of cells. They do have the potential to make a person, and it can therefore be argued that they must be treated as one would treat any other person. That is, in very direct terms, they must not be killed. This is a debate that is not likely to be settled in the near future.

Is there another way to move ahead, avoiding this controversy? One option resides in another category of stem cells, the so-called adult stem cells. It turns out that some cells with stem-cell properties persist even in the adult. We've already mentioned blood stem cells in bone marrow. But these stem cells are only capable of making blood. There are also stem cells in the adult brain that continue to divide throughout life, generating new neurons. In theory these cells could be harvested and used to repair a damaged brain. Furthermore, it has been reported that some of these stem cells found in one organ are also able to produce cell types from other organs. For example, it has been claimed that brain stem cells are capable of making blood. But, despite a flurry of articles in the most influential journals, many of these assertions have remained controversial in the scientific community, because some of the experimental results have been impossible to reproduce in other laboratories. It is generally agreed that adult stem cells are less powerful than embryonic stem cells. They are more limited in their potential to make different cell types. And some of these adult stem cells present considerable technical challenges. For example, the brain stem cells are located deep within the brain, essentially inaccessible. We can't get at them without causing significant brain damage in the process. So, while for some organs there are indeed residual stem cells even in the adult, there remain important questions concerning the power of these cells and their availability. They offer both promise and problems. Research in this area needs to be pushed forward, but it isn't clear that adult stem cells will provide the miracle stem-cell therapies that we hope to eventually achieve.

But now there is a third option. It offers the power of the embryonic stem cell but avoids all of the ethical controversy. This is clearly the option of choice. It is based on new developments that we could not have imagined just a few years ago. It involves, in essence, taking adult cells and reprogramming them to become embryonic stem cells. This option is near ideal in many respects. It would allow us to make the stem cells from the patient, perhaps from a skin sample, so the cells are an exact genetic match, and therefore when they are reintroduced into the patient for the sake of therapy there would be no chance of the immune system rejecting them. And even though these stem cells are derived from the cells of the adult, they have the full power and developmental potential of stem cells made from embryos.

To better understand how this new approach works, we need to take a brief historical trip that reviews some of the highlights of our discovery of how genes control the developmental potential of cells. We now know that a typical mammalian cell uses over ten thousand genes. Indeed, sen-

sitive modern assays can detect some level of activity, RNA
being made from about half of the twenty-five thousand
genes present in the genome. This suggests that there might
be an incredibly complex gene-expression code that defines
the identity of a cell. Thousands of genes might be involved.
Scientists found, however, that there is an astonishing underlying simplicity, with a very small number of genes playing a master role in defining cell types. This principle was
eventually exploited to make the conversion of human
adult cells into stem cells possible. Perhaps surprisingly, the
story begins with insects.

About a hundred years ago, geneticists working with
fruit flies made a series of shocking discoveries. They were
mutating the genes of the flies, with chemicals and radia-
tion, and then observing the resulting effects. A hundred
years ago we didn't even know that genes were made of
DNA. We didn't really have any idea how genes worked.
We did know, however, that they existed, that they were on
chromosomes, and that they determined the traits of organisms. Mendel didn't need to know the chemical nature of
genes to study their inheritance patterns in peas. Anyway,
when the fruit-fly geneticists mutated certain genes, by
chance, they observed some truly amazing flies. One mutation, for example, caused embryonic cells that would normally make the antennae to instead make an extra set of
legs that now protruded from the head where the antennae
were supposed to be! Imagine a fly with legs coming out of
its head. This gene was appropriately named *antennapedia*
("pedi," denoting foot or leg, so the name means antenna-
leg). Another mutation produced a different crazy-looking
fly, in this case with a complete extra set of wings, so the fly
had four wings instead of the normal two. There were

about half a dozen of these genes that, when mutated, dramatically altered development, causing cells that would usually make one structure instead make something quite different.

These genes appeared to be "master switch" genetic regulators of development. They were like switches on a train track. Throw the switch the wrong way and the train ends up going to Florida instead of New York. Mutate the gene the wrong way and you get a leg instead of an antenna. In both cases a seemingly simple flip of a switch could have an enormous effect on the end result. The more you think about it, the more shocking this sort of result is. After all, the leg and the antenna are both quite complex structures, with many different individual cell types, expressing thousands of genes. So how on earth could just one gene have such a dramatic impact?

When these genes were eventually sequenced, many years later, it was discovered that they encoded proteins called transcription factors. This provided a partial explanation of their developmental power. These genes encoded proteins that regulated the expression patterns of other genes. And some of the genes they regulated were also transcription factors, which could in turn regulate still more genes. The result could be a genetic cascade, with one gene at the top eventually controlling the expression levels of hundreds or even thousands of "downstream" genes.

Another way to view the process is as a genetic pyramid. At the top of the pyramid is a master-switch genetic regulator, capable of controlling the expression patterns of a set of genes located beneath it on the pyramid. Some of these genes can reach further down on the pyramid, regulating the expression of additional sets of genes. This process continues down until the bottom is reached, where the last set of genes is located; these genes drive the formation of the final differentiated state.

When these master-switch genes were sequenced we not only discovered that they encode transcription factors, we also found that they share a common block of sequence, 180 bases long, called the homeobox. The half a dozen or so genes of this type in the fruit fly all have a homeobox, although it differs somewhat in base sequence from one gene to the next. The rest of the gene sequences are generally very different, but the homeoboxes are quite similar. In other words, these genes make up what is called a gene family, because they all share a related sequence, in this case the homeobox. In evolutionary terms this means that if you go back far enough in evolutionary time there was an ancestral organism that had just one copy of this type of gene. This copy was then duplicated to give two copies, and then to give many copies, and these copies then diverged somewhat in sequence, through mutation, and assumed different functions. Such gene families are extremely common in the organisms of today, more the rule than the exception, and this process of gene duplication followed by divergence is thought to be a very common tool of evolution, allowing it to start with a gene of one function, and to modify copies of it to assume new functions. This is much more efficient than starting with random DNA sequence, with no function, and evolving it into something useful.

But fruit flies, of course, aren't the only organisms that have these master-switch homeobox genes. Indeed, all mammals, including mouse and man, also have them, and their master-switch genes also include the 180-base-pair homeobox. It thus appears that the same class of genes that can regulate formation of the fruit fly can also drive development of mammals. Some very interesting genetic studies have further confirmed the close functional relationships of the fly and mammalian master-switch genes. Indeed, a number of experiments have shown that the mammalian and fly homeobox genes are often functionally interchangeable! For example, if you take the mammalian homeobox gene that is most closely related to the fruit fly *antennapedia* gene, and you mutate it the same way the fly gene was mutated when the antennae were converted to legs, and you then insert this mutant mammalian gene into the fly, you find that, astonishingly, the antennae are again converted to legs. That is, it doesn't matter if you have a mutant copy of the fly gene, or a mutant copy of the corresponding mouse (or human) gene, you get the same result, legs on the head instead of antennae. This doesn't work in all cases, but the fact that it can ever work at all is quite surprising.

One might ask, If you place the mutated mouse *antennapedia* gene into the fly, do you get little mouse legs coming out of the fly head? The answer is no, you still get fly legs. And the reason is that all of the other genes of the fly remain fly genes. That is, the genes beneath the *antennapedia* gene in the genetic pyramid of leg development remain fly genes. So, while the mouse gene in the fly is indeed capable of flipping the switch, initiating the genetic cascade, starting the genetic program, the genes that it has to work with within the context of the fly are all fly genes, so the end result is a fly structure and not a mouse structure.

Another dramatic example of a master-switch gene is the *eyeless* gene in flies. As you might expect from the name, when this gene is mutated in the fly no eyes are formed. The results of extensive genetic experiments show that in the fly this gene is, amazingly, capable of initiating the complete genetic program of eye development. That is, this is the gene at the top of the genetic pyramid for eye development. If you inactivate this gene by mutation, then the eye program does not initiate, and no eyes form. But more striking, if you genetically activate this gene in places where it would not normally express, then you get extra sets of eyes. For example, if you turn this gene on in the wings, then eyes form on the wings, and if you activate this gene in the legs, or antennae, then eyes form at these unexpected positions. Indeed, by turning the gene on in multiple positions you can make a fly that has eyes not only where they should be, but also on the wings, the antennae, and the legs. A crazy-looking fly, with lots of eyes.

As for the fly with legs in the wrong place, this fly with extra eyes is the more remarkable the more you think about it. The fly eye, referred to as a compound eye, is a very complex structure, consisting of lots of little eye units, each having its own lens and set of photoreceptor cells for detecting light. It takes thousands of genes to make a fly eye, yet this one gene, *eyeless,* starts the genetic program and in the end gives a complete eye.

And then came the real surprise. Like the *antennapedia* gene, the fly *eyeless* gene was functionally interchangeable with its mammalian counterpart, in this case named *Pax6.* That is, one could insert the mammalian gene into the fly, making what is referred to as a transgenic fly, and if you then activated this gene in some strange position during development, like the wing or the leg, you would once again get an extra set of eyes in that unexpected spot. And, again you would get fly eyes, not mammalian eyes, because although the mammalian gene could work in the fly and activate the genetic program of eye development, the genetic program present in the fly is a fly program, and not a mammalian one. This business of making extra eyes in the fly could be accomplished with either the mouse or the human *Pax6* gene.

This discovery was particularly unexpected because it had been thought that the insect eye and the mammalian eye were not evolutionarily related. There are so many fundamental structural and operational differences between them that biologists had thought this represented an example of convergent evolution—two structures evolving not from a common ancestor, but instead independently, although in a similar direction, to give a light-detection organ. But the observed similarities in the genetic programs of insect and mammalian eye development, with the *eyeless/Pax6* genes functioning interchangeably to initiate the genetic program, argued very strongly for an evolutionary relationship including a common ancestor with a primitive eye. That is, at some distant time in the past, we think around a few hundred million years ago, there was an organism with a primitive light-detection organ. Already, even back then, the genetic program that was used to make this primitive eye included the use of the *eyeless/Pax6* ancestor gene as a master regulator. The progeny of this primordial beast evolved in many different directions, but this genetic program for eye production remained, at least in its basic form.

This makes sense if you think about it. If you mutated a gene such as *eyeless,* which was a key regulator of eye formation, then you would lose your eyes, and this would be very bad for survival, making it less likely that the mutation would be passed on. Such genes at the top of genetic cascades of development are very well conserved during evolution just because of their extreme importance. If you mess with them you dramatically change the resulting organism, losing antennae, losing legs, or losing eyes, all of which are essential structures. Instead, evolution more often produces changes of genes responsible for the details of a structure, because these give more acceptable results. So once a genetic program for generating an essential structure, like the eyes, is established, then it is conserved, at least at the higher levels of the genetic program, for very long evolutionary periods, even the hundreds of millions of years separating fruit flies from people.

Further confirming the common evolutionary origin of eyes, scientists have found that the squid eye uses a similar genetic program of formation, with a gene quite similar to *Pax6* again at the top of the genetic pyramid. The squid *Pax6* gene is also capable of driving extra eye formation in flies, just like the mammalian gene. So we now believe that a common ancestor, living about half a billion years ago, with primitive eyes, gave rise to the insects, the mollusks (including squids), and the mammals of today. And some aspects of the genetic program of the development of that primitive eye of five hundred million years ago persist today, and are shared by these very diverse organisms.

These fruit-fly examples show how single genes can initiate extensive programs of development, driving the formation of complex organs and structures, like legs and eyes. It is therefore perhaps not surprising that there are also genes that are capable of dramatically altering the developmental state of individual cell types. One of the first-discovered extraordinary examples of this is the *MyoD* gene, which can change many different cell types into muscle. This was quite a striking finding. For example, the addition of an active *MyoD* gene to an adipoblast, a fat cell, would convert it into a muscle cell. This gene is accomplishing something similar to the *Pax6* and *antennapedia* genes, only on a smaller scale. Instead of driving the formation of an entire structure or organ, which involves many different cell types, the *MyoD* gene is converting one type of cell into another. This is still a considerable accomplishment, requiring the *MyoD* gene to start within the cell a genetic cascade that involves large numbers of genes. Someday we might be able to rub a *MyoD* cocktail on our fat tummies and develop six-pack abs, without the exercise.

Now we return to the stem-cell story. We've seen that it is possible to express a master-switch genetic regulator in fruit flies, and to drive the development of groups of cells into legs on the head instead of antennae. We've also discussed how the *MyoD* gene in mammals can initiate a genetic program that converts many cell types, including fat, into muscle. These kinds of observations set the stage for an even more remarkable set of discoveries. Could it be possible to find a gene or genes that could take an adult, differentiated cell, and turn back the clock, reverse its development, and transform it back into an embryonic cell? At first glance this would seem an impossible fantasy. Incredibly, however, it has now been done.

As you might expect, it wasn't very easy. Scientists had studied embryonic stem cells carefully, and knew all of the genes they expressed. But they expressed a lot of genes, over ten thousand, so which of these genes might be the secret answer to making adult cells revert to stem cells? It was like looking for a needle in a haystack. But scientists had some clues. We knew that one class of genes, those encoding transcription factors, would provide the best candidates. These are the genes that regulate other genes and are capable of initiating genetic programs. The fruit-fly master-switch regulators and the mammalian *MyoD* gene all encoded transcription factors. But even when we limited the candidate genes to those encoding transcription factors we were left with a list of over a thousand.

It was possible to trim the list further by comparing the genes expressed in the embryonic stem cells to those expressed by other cell types. It turns out that for some transcription-factor genes we find that their expression is mostly limited to stem cells. They are rarely expressed anywhere else, suggesting that they might be particularly good candidates for driving the stem-cell genetic program, for making adult differentiated cells into stem cells. But even after applying this filter there were a few dozen genes left. And when each of these genes was tried individually, nothing happened. The adult cells stayed adult cells and were not converted into stem cells.

This suggested that perhaps this strategy was not going to work. We just would not be able to reverse development and turn adult cells back into embryonic stem cells. Or that perhaps it would be possible, but would require a combination of genes instead of a single gene. While the conversion of a fat cell into a muscle cell only required one gene, *MyoD,* maybe the conversion of an adult cell into an embryonic stem cell was more difficult, and would require a combination of two, three, four, or more genes. So, many different combinations were tried.

Amazingly, it worked! Shinya Yamanaka and his colleagues at Kyoto University found that when a magic mix of four active genes, including *c-myc, sox2, klf4,* and *oct4,* were introduced into an adult mouse cell it was possible to transform the adult cell into a stem cell. And this new stem cell was essentially indistinguishable from one derived from an early embryo. It was possible to treat it in different ways and to induce it to form many distinct types of adult differentiated cells, including those from the pancreas, heart, and brain. Furthermore, these adult-derived stem cells passed the ultimate test: by using the embryologic methods discussed in the following chapter it was possible to turn one of these cells into a complete individual, a mouse, with all cell types present and normal. In the terms of the developmental biologist, these cells are totipotent: able to give rise to all of the different kinds of cells found in an adult.

The discovery of a method to turn differentiated adult cells into stem cells completely changed the landscape. Again, it is now possible to think of taking a patient, removing some skin cells, turning these into stem cells, and then using these stem cells for a host of different therapeutic applications. These cells eliminate the ethical controversies associated with stem cells derived from embryos, and because they are actually derived from the patient, they are a perfect genetic match and would not be rejected by the patient's immune system when reintroduced. These adult-derived embryonic stem cells offer enormous promise. Their power is unmistakable, they will in time save countless lives, and one can confidently predict that Yamanaka will win a Nobel Prize for his discovery.

There are, however, still many mysteries associated with these cells. For example, the magic cocktail of four genes only works in a fraction of cells. When the procedure was first described it was extremely inefficient. If ten thousand skin cells from an adult were treated with these genes, only a few, maybe ten at most, would actually become stem cells. More recent developments have dramatically improved the efficiency, with numbers as high as one in ten reported. We still don't understand why only some cells are converted. Nevertheless, the efficiency is still high enough to be of use. Even a single stem cell can be grown in culture under the right conditions to give millions, or billions, of stem cells. They continue to divide for a very long time. So it will indeed be possible to turn skin cells into stem cells, albeit with low efficiency, and then to grow the stem cells in culture to expand their numbers, and to use them for disease therapy.

But what does all of this have to do with making designer-genes babies? It turns out that we are learning how to turn stem cells into eggs and sperm. That is, remarkably, it might be possible within a few years to take stem cells and to convert them into eggs and sperm by using the right set of culture conditions. This could have enormous implications in terms of offering much higher numbers of embryos to screen for the presence of desired gene combinations. As noted previously, men produce huge numbers of sperm, but women produce relatively few eggs at a time. Even with multiple rounds of hormone treatments it is possible to retrieve at best a few dozen eggs from a woman. This means there can only be a few dozen embryos to choose among. The use of adult-derived stem cells offers the potential to produce unlimited numbers of embryos for screening. For example, a tiny amount of skin is taken from a woman, and a few of those skin cells are turned into stem cells by treatment with the four-gene mix. These few cells are amplified by growth in cell culture to essentially unlimited numbers of stem cells, which would then be turned into eggs by changing the culture conditions. We haven't yet worked out all of the details of how to turn stem cells into functional eggs, but much progress has been made, now allowing us to turn stem cells into the precursors of eggs. It is likely that in the future we will figure out how to turn these precursors into eggs, which would then be fertilized by sperm from the father, using in vitro fertilization techniques. This would give an unlimited numbers of embryos, which would then be genetically assayed by removing single cells and performing whole genome DNA sequencing. The DNA sequences would reveal the predicted character combinations of the potential progeny, such as health, longevity, appearance, intelligence, and so on. The parents could choose the embryo with the best combination of genes, according to their criteria, and this embryo would be implanted into the uterus of the mother and be born nine months later as a designer-gene baby.

11

MODIFYING GENES

The latest genetic-engineering technology allows us to use stem cells as the starting material to produce individuals with any desired version of a gene. With this approach we can go further than just sorting and selecting from a batch of embryos. Even after examining a large number of embryos the favored gene combination may not be found. In some cases neither parent will carry the sought-after form of a particular gene. With genetic-engineering tools, however, it is possible to take a more active role, and to actually modify genes at will.

In 2007 the Nobel Prize in Medicine or Physiology was awarded to Mario Capecchi, Sir Martin Evans, and Oliver Smithies for the development of this radical new technology. It can be used to accomplish many different types of genetic modifications. It can be used to delete a gene entirely, or to add a new gene that might have been synthesized from scratch in the test tube. Or it can be used to change a gene that is already present in the cell. This change can be major (for example, swapping one large block of DNA sequence for another), or it can be extremely subtle (such as modifying a single base of a gene, perhaps turning a C into an A). This new technology gives us total control over the genetic content of a cell. The genetic-engineering procedures involved are complex, currently requiring months of work in the laboratory to accomplish. It is not as simple as pushing a button. But the results are stunning.

There are three steps. First, the desired version of the gene is made in the test tube. Second, this synthetic gene is introduced into stem cells grown in a plastic dish in an incubator. Third, and perhaps most surprising, a correctly genetically engineered stem cell is used to make a person, carrying the engineered genetic modification in every cell of his or her body. This might sound more like science fiction than science, but these same procedures are routinely used today in research laboratories around the world to genetically modify mice. There is every reason to suppose that this technology would also work for people.

To elaborate: The desired gene changes are created in a test tube using standard molecular-biology techniques. For example, the artificial gene could simply be generated de novo by a DNA-synthesis machine. We just type in the desired base sequence and the machine produces large numbers of the specified DNA molecules. It is possible to synthesize not just single genes, but large collections of genes using such machines. Indeed, the entire genome of a simple bacterium, with all of its many genes, has been machine made. Alternatively, we could take a purified copy of the gene of interest and introduce the desired changes manually, using recombinant-DNA strategies developed in the 1970s. One can purchase enzymes that are used to cut DNA molecules at specific places, and other enzymes for pasting pieces of DNA together. Still other enzyme combinations can be used to change single bases of genes being manipulated in the test tube. But as DNA-synthesis machines continue to improve, with increased accuracy and reduced cost, it is likely that they will replace these older, more cumbersome approaches to making modified versions of genes.

The second step is to genetically engineer the stem cells. The challenge is to replace the gene in the stem cell with the new version that now sits in a test tube. The copies of the modified gene are inserted into the cells by electroporation, which means we zap the cells with a jolt of electricity, which opens up holes in the membranes, allowing the new DNA molecules with the customized genes to enter. The next stage of the process occurs within the stem cell and is not terribly well understood. Somehow the DNA introduced into the cell finds the gene to be changed, the one originally present in the cell, and lines up with it, side by side. This is not a trivial accomplishment, since there are billions of bases of DNA within the cell, yet by unknown mechanisms the introduced DNA does find its partner, the few thousands of bases of the gene to be modified. Then DNA recombination actually replaces the original gene with the altered version, thereby accomplishing the desired genetic engineering.

As discussed previously, the process of recombination is a natural part of meiosis, the special cell divisions that generate the gametes, the eggs and the sperm. Recombination normally mixes up the gene combinations present, and is responsible for the different sets of genes normally found in multiple children of a single pair of parents. Every person is diploid, with two copies of every gene, but in the business of making gametes we must reduce the gene content to one copy of each gene, so when the gametes unite at fertilization the normal state of diploidy will be restored. Recombination between pairs of chromosomes helps to jumble the genetic content of the different gametes, so they each carry different sets of gene variants from the parent. Scientists can use this natural recombination capacity present in many cell types to engineer changes in genes.

The genetic engineering technology used to replace a gene or genes is indeed powerful, but very inefficient. It only works in about one cell in a million. So if we start with ten million stem cells, a typical number for these types of experiments, we only end up with ten cells that are properly genetically modified. It is amazing that it works at all, but the fact that it works so rarely creates challenges for the scientist, because it is necessary to search through the ten million cells to find the ten that have the desired genetic modification. A series of screening procedures are used to find those rare cells where everything went right.

There are three possible outcomes when the DNA with the gene modification is introduced into the stem cells. The most common is the failure of the modified DNA to integrate into the stem-cell DNA at all. It simply floats around in the cell and is eventually degraded and lost. To screen against these cells we typically tag the modified gene with the presence of an antibiotic-resistance gene, so when it is joined into the cell DNA it carries with it the property of antibiotic resistance. Then by simply growing the cells in the presence of the antibiotic we kill the stem cells that did not incorporate any of the new DNA. This is a very useful selection that removes all but a few hundred cells. Another common outcome is that the introduced DNA integrates into the wrong place in the stem-cell DNA, and does not accomplish the desired gene modification. This is the result of a very poorly understood process called illegitimate recombination. For some reason the introduced DNA goes into some random position in the cell DNA, instead of the gene we intended to modify. Ninety to ninety-nine percent of the cells that survive antibiotic selection have usually suffered such indiscriminate integration events. And of course the final and desired outcome is the replacement of the original gene by the modified version.

It is difficult to distinguish the stem cells with the correct gene modification from those that have a random integration of the introduced DNA. Both types of cells carry the antibiotic-resistance gene and survive the antibiotic selection step. The stem-cell screening procedure therefore requires a laborious analysis of each surviving cell. The cells are grown on a plastic dish, well separated from one another. Each cell divides, to give two cells, which divide again, to give four cells, and so on. After a few days of growth there is a clump of cells where each individual cell started. Scientists call these clumps colonies, and they are also referred to as clones. Each colony represents the offspring of a single original cell, and because DNA replication makes daughter cells that are genetically identical to the parent, all of the cells of the colony are genetic copies, or clones, of the original cell. The hundreds of colonies are individually "picked," or removed from the plastic dish, and divided into two parts. One piece of each colony is used for DNA analysis, to determine if the proper genetic modification took place. The other part of the colony is frozen in liquid nitrogen, to preserve it in a state of suspended animation while the DNA screening is carried out. The colonies will stay preserved indefinitely until they are needed, at which time we simply thaw those stem cells that were found to be correctly genetically modified, return them to the growth medium, and use them as we wish.

The part of the stem-cell colony taken for screening has only a small number of cells and a tiny amount of DNA. The DNA analysis of the clones therefore usually employs a technology called the polymerase chain reaction, or PCR. This is a remarkable tool that allows the characterization of extremely small samples of DNA. When forensic scientists are examining the DNA from a single hair, for example, to determine if a suspect might be responsible for a crime, they are using PCR. The procedure employs a powerful chain-reaction strategy, as the name suggests, to amplify DNA. In essence, very short pieces of DNA bracketing the region of DNA of interest are first synthesized by a machine. These pieces, called primers, are then used to prime the synthesis of the desired DNA region by an enzyme called DNA polymerase. The result is that a copy of the sequence of interest is made, so there are now two copies instead of the original one. Then the whole process is repeated, with the primers now driving the replication of not only the original DNA sequence, but also the copies. This is repeated again, over and over, each time doubling the number of copies present. After ten cycles about a thousand copies have been made, and after twenty cycles there are a million copies. By using PCR it is possible to start with very small amounts of material, such as a few cells from a colony, and to make enough DNA to allow a sequence analysis that will tell us if the cells have the desired gene modification or not.

Kary Mullis received the Nobel Prize in chemistry for his development of PCR. The prize was a testament to the importance of this technique in modern molecular biology. Mullis is quite an eccentric figure in science. He earned his Ph.D. from the University of California, Berkeley, in 1972, and has been quoted as saying that during this period he "took plenty of LSD," and that this "was certainly much more important than any courses I ever took" (*California Monthly,* September 1994). After winning his Nobel Prize he would sometimes stray from the expected topic during his lectures. One particularly famous example was a presentation at a medical society meeting in Toledo, Spain, in April 1994, where his only slides were of naked women he had photographed (*The New York Times,* September 15, 1998). But we also divert from the topic at hand.

The third and final step is to take the genetically modified stem cells and turn them into babies. This is a truly amazing transformation if you think about it. How can clumps of cells grown in a plastic dish in an incubator be turned into designer-genes babies? Technically it turns out to be not all that difficult. The required procedures are used routinely in mice, and would presumably also work for people. The stem cells would simply be added back to an early embryo. Such work is normally accomplished with some very fine, delicate glassware and with micromanipulators, which allow the scientist to make extremely precise movements on a very small scale. Once the stem cells were back in a proper embryo environment they would proceed to make a baby, which would be born normally, nine months later.

In summary, the basic scheme for the production of designed genetic modifications in people is relatively straightforward. First, the desired gene is made in a test tube, perhaps by just entering the DNA sequence into a machine. Second, the modified gene is used to genetically engineer stem cells. This is currently an inefficient process, involving the screening of many cells to find those with the sought-after gene change. And third, the stem cells are then used to make a genetically engineered person.

GENETICALLY ENGINEERED MICE

As mentioned above, these exact procedures have already been used, thousands of times, to make mice with a desired set of genes. In the world of the mouse geneticist these techniques are referred to as gene targeting, because they allow the scientist to target specific modifications to any desired gene. Scientists use gene targeting to study gene function. Perhaps, for example, one suspects that a particular version of a gene will make a person susceptible to cancer. It is known that mutations in the *BRCA1* and *BRCA2* genes have been associated with a very high risk of breast cancer in women. To study the specific roles of different versions of genes in cancer, scientists can genetically engineer mice and then observe if the modified mouse does turn out to be more prone to cancer. Or perhaps the scientist thinks a gene is especially important in the formation of the kidney. If so, one would predict that this process would not proceed normally with the gene absent. A targeted deletion of the gene can be engineered, thereby entirely removing its function, and one can study the resulting mice and see if kidney development is indeed disturbed. Mutant genes also underpin a host of genetic diseases, and gene-targeted mice can be used to model these diseases to better understand the disease processes and to develop improved therapies. It is easy to see why gene targeting has been a tremendously useful tool for a variety of scientists studying a huge range of different topics.

A TECHNICAL ASIDE

We have extensive experience in converting genetically engineered stem cells into mice. For gene targeting in mice the standard procedure is to introduce the genetically altered stem cells into an early embryo, a blastocyst. Once the stem cells are within the blastocyst they join with the embryonic cells already present, the inner cell mass. This results in the birth of a chimera, with some cells that are not modified, from the host embryo, and some cells that are genetically altered, from the introduced stem cells. This is great for mice, where we have methods for breeding the chimeric mice to make offspring that carry the genetic modification in all cells. But for people we don't want to make chimeras, we want to directly make individuals that are uniformly genetically modified, in every cell of the body. The mouse-model system has shown us two strategies that allow this to be accomplished.

One method is to use a host blastocyst that is tetraploid—that is, it has four copies of every gene instead of the normal two. It has been found that the tetraploid cells can only make support structures, such as the placenta, and cannot contribute to the embryo itself. In this case all of the cells of the resulting baby would be derived from the inserted genetically modified stem cells, which are diploid. Such tetraploid host embryos are easily made by allowing a one-cell fertilized egg to divide once, and then fusing the resulting two cells back together into one cell, which would now have four copies of every gene. This tetraploid embryo can then be cultured and will proceed to make an apparently normal blastocyst, which can be used for genetically engineered stem-cell injection.

Another approach is even simpler. The genetically modified stem cells are inserted into an earlier stage embryo, when it has only two or four cells—well before the blastocyst stage. It has been shown that in this case the resulting mouse is entirely derived from the added stem cells.

In summary, our extensive experience with the mouse-model system shows that it is indeed possible to genetically modify stem cells and then to make individuals from them.

THE SOURCE OF THE STEM CELLS

Where would the stem cells used to make designer-genes children come from? As discussed in the previous chapter, one excellent source of stem cells is a five-day-old embryo, the blastocyst. A likely scenario, therefore, would be the following. The prospective parents would produce a batch of early, eight-cell embryos, and single cells would be removed for DNA-sequence analysis. The embryo with the best gene combination would then be grown a bit longer in an incubator to the blastocyst stage and used to make stem cells, which would then undergo final genetic fine-tuning. These modified stem cells would then be used to make the child.

As we discussed in the previous chapter it is also now possible to turn skin cells into stem cells. This could facilitate the creation of large numbers of eggs from the prospective mother, and provide large batches of embryos to be screened. But it also raises another, perhaps more sinister possibility, stem-cell cloning.

STEM-CELL CLONING

The announcement of the birth of Dolly the sheep surprised the world in 1997. This was the first example of the making of a clone, or a genetic copy, of an adult mammal. Dolly was made using a nuclear-transfer procedure. First, the nucleus of an egg from a Scottish blackface sheep was removed and replaced with the nucleus from a cell from the mammary glands of an adult Finn Dorset white sheep. Then the egg, with the new nucleus, was stimulated to start dividing. This can be done with a pulse of electricity or chemically. After a few days of growth in an incubator the resulting embryo was surgically placed into the uterus of a Scottish blackface sheep, eventually resulting in the birth of a lamb with a white face. She was named Dolly after the well-endowed singer Dolly Parton, because she was made with the nucleus of a mammary-gland cell.

It is amazing that this procedure actually works and makes clones. Of course, every cell of the body has the same complete set of genes, so this isn't a problem. But the mammary-gland cell would be using its mammary-gland genes and not the same genes that an embryo would use. The wrong genes would be active. The key is that the cytoplasm of the egg can somehow turn off the mammary-gland genes of the transferred nucleus, and turn on the embryo genes. This is called cytoplasmic reprogramming, and is not well understood. Furthermore, it is extremely inefficient and imperfect. It was necessary to do 277 nuclear transfers, into 277 eggs, to get one surviving lamb. All of the rest of

the embryos died somewhere along the way. The reprogramming process didn't work well enough for them to
make a normal embryo, and lamb. And even Dolly wasn't
completely normal. She died at age six, while most sheep
will live eleven or twelve years. There is some evidence that
she aged prematurely. The ends of her chromosomes, called
the telomeres, were too short, because she was started with
a six-year-old nucleus from the mammary gland of an adult.
These ends normally shorten with age in all of the cells, except those that make eggs and sperm. Nevertheless, subsequent work has shown that it is possible to use nuclear
transfer to clone cows, pigs, mice, goats, and many other
mammals. For a price you can even have a cloned copy
made of your pet cat.

But nuclear-transfer cloning doesn't really work properly. Even the best-looking clones aren't entirely normal
when carefully examined at the molecular level. In addition
to problems with the ends of chromosomes being short, the
gene-usage patterns aren't quite right either. The nuclear-
transfer technology is a flawed approach, which generates
embryos of which 99 percent are defective and don't survive
to birth, and a few that do survive, but are still imperfect.
Obviously this is a technology to be avoided.

The latest developments in the stem-cell field, however,
dramatically change the situation. There is now another option, cloning through stem cells. We have acquired a deeper
understanding of the genetic reprogramming process. We
can now make the equivalent of early embryonic stem cells
from the skin of an adult, and in every measurable respect
these cells appear normal. And, at least with mice, we know
how to take embryonic stem cells and turn them into perfectly normal newborns. We simply place the cells in with

an early embryo, which will provide the placenta and other support structures. After growing a few days in an incubator the embryo is implanted into the uterus of the mother, where the stem cells proceed to make a healthy newborn.

It therefore appears that science has marched forward, and the technological objections to human cloning might have been removed. But, of course, the moral issues remain, and are now even more complicated. Suppose that a person would like to be cloned, but with genetic improvements. "Just like me, but a little smarter and better looking." While nuclear-transfer cloning was viewed as a method for making identical genetic copies of individuals, stem-cell cloning is easily adapted to the production of modified versions of people. Simply take the stem cells, genetically engineer them, and then use them to make semi-clones, which would be designed perhaps to be a little healthier and happier. One can imagine a possible scenario in the future when a couple would choose to raise two children that are genetically enhanced semi-clones of the father and mother. These improved copies in turn would be further improved in following generations. People with greater financial resources could make more copies, with greater enhancements.

12

BUT IS IT MORAL?

It is clear that in time we will have the scientific means to select the genes of our children. Over the coming years we will acquire the knowledge of which gene combinations will result in which traits. This means that we will be able to select the features of our children, ranging from appearance to mental health to physical ability and intelligence.

But should we use this new power? Will it be a good thing, allowing us to eliminate genetic disease, and to ensure the best possible genetic endowment for our children? Or is it an evil, resulting in the murder of countless unborn embryos that don't meet our lofty genetic expectations?

According to one view we may be entering an age where our genome, our DNA, is able to manipulate its own evolution. It uses us, the people that this DNA codes for, as the instrument of change.

Does human intervention in the shaping of its own

species represent an intrusion into a domain rightly reserved for a higher power? Should the human species be allowed to play God with its genetic destiny? Is this new genetic technology as menacing as the knowledge that allows us to make an atom bomb? Is it something dangerous, and indeed threatening, to the existence of the human species as we know it? Is it a technology that demands extremely tight controls? Should governments outlaw it? Or is it an effective new tool for good that will result in a dizzying spiral of evolution, with each new generation more intelligent, and therefore better able to design the genetic content of the following generation?

WHEN IS THE DESTRUCTION OF HUMAN EMBRYOS MURDER?

The genetic manipulations described in this book for the creation of babies with desired gene combinations would result in the destruction of human embryos. With present technology this would appear unavoidable, although one cannot rule out technological advances in the future that might eliminate this requirement, as discussed in chapter 17. Is this technology, therefore, in its current form, morally unacceptable because it requires the destruction of human embryos? Are we considering committing the murder of some human beings to improve the genetic makeup of others? Is this a clear example of how the ends do not justify the means?

There are no certain answers to these questions, only different points of view. The procedures described in this book are extensions of current in vitro fertilization techniques. As we discussed previously, these in vitro fertilization procedures, as used today, create many embryos in the test tube by combining sperm and eggs, and only a few of these embryos are ever born as children. Most are discarded or frozen indefinitely. That is, we are already sacrificing human embryos.

Is this evil? While it does kill human embryos, it also allows otherwise impossible births. More than three million total children, worldwide, have been born through in vitro fertilization over the years since Louise Brown. Without this procedure these children would not exist. So, from one perspective, this procedure has allowed the creation of otherwise impossible human life. There are currently about a hundred thousand in vitro fertilization births per year in the United States, a massive number approximating 1 percent of all births. If we were to outlaw in vitro fertilization we would be eliminating the existence of these hundred thousand new children each year.

One could argue, however, that in vitro fertilization, which helps otherwise infertile couples conceive, is fundamentally different from the procedures described in this book. It helps to create new life, while prenatal genetic screening, as used today, is more of a negative screen, eliminating life that has already been created. And the methods used to make genetically engineered children are primarily a selection scheme, determining which embryos will live and which will die. In these cases we are not sacrificing the lives of some embryos to create the lives of others, but using some embryos to improve the genetic constitution of our offspring. The embryos with the best gene combinations survive, while the rest are discarded. Can we justify using these techniques when the question is not one of existence or nonexistence for our offspring, as it is in the case for the infertile couple? When, rather, it is deciding if our children, which could be born anyway, will have the most desirable genes or not?

ONTOGENY RECAPITULATES PHYLOGENY

An important principle in biology is that ontogeny recapitulates phylogeny. This means that the development of the individual copies the evolutionary history of the species. For example, it is fairly common knowledge that the early human embryo has a tail, just as the evolutionary ancestors of humans had tails. And human embryos have gill-like structures, called branchial arches, that are reminiscent of the gills known to have been present in the ancient sea-dwelling ancestors of all land life.

The kidney provides a dramatic example of this principle. In the early human embryo a pair of very primitive kidneys are the first to form. This early kidney type, named the pronephros, is simple in construction and resembles the kidneys that function in embryonic fish and the tadpoles of frogs. These early nonfunctional human kidneys then deteriorate and are replaced by somewhat more sophisticated versions, which resemble the kidney type observed in adult fish and amphibians. In female humans these kidneys also deteriorate, while in males they are incorporated into the reproductive tract. Then the third and final kidney type, the metanephros, forms. These are the kidneys found in essentially all land vertebrates, designed to cope with the challenges of limited water supply and able to retain more fluid, thereby reducing water loss.

Why does the human embryo bother to make these early primitive kidneys before making the final, functional kidney type that it will keep and use? It seems a somewhat wasteful approach, and not terribly intelligent. But, at least in some cases, it appears that evolution works best by adding on rather than replacing. The embryo keeps making primitive kidneys because of an ancient memory telling it to do so, even though it doesn't need those primitive forms anymore. Indeed, the early pronephros is completely nonfunctional. It is further proof of evolution, as why else would you find these primitive structures in human embryos? If you were designing a human from scratch you wouldn't likely want to waste your energy making ancient fishlike and amphibian structures, only to then discard them.

One can also use this principle to argue that the early human embryo is not really very human at all. It certainly doesn't look human. Embryologists have long observed that it is very difficult to distinguish early human embryos from those of other mammals, or even from those of chickens or fish. Should we consider that when the human embryo is just developing, and has a very small brain, a tail, gills, and a fish- or frog-type kidney, that this tiny organism is really more like a fish than a human? And, indeed, the embryos that would be sacrificed in the procedures described in this book are much earlier still, just a tiny ball of cells, with no nervous system at all, no beating heart, resembling the most primitive multicellular (metazoan) life forms that first appeared on the surface of earth.

Of course the thing that separates the early human embryo from a primitive metazoan or a fish is the potential to eventually form a person. So, while it might not look like much, and might have a structure and a set of organs that resemble a fish's, it is indeed set apart from these other life forms by its capacity to make a human being.

WHEN IS AN EMBRYO A PERSON?

So, when is an embryo a person, with full rights, including the all-important right to exist and not be killed? This is the key question. The answer determines whether the work described in this book should be allowed to proceed.

The Catholic Church gives a very clear answer to this question. According to Catholic doctrine the embryo becomes a full person, with all rights, at the moment of conception, when the egg and sperm come together. In some respects this is an attractive and theoretically clean answer. It is certainly difficult to imagine picking a point sooner than this. While gametes, the eggs and sperm, could be argued to have the potential to make people, needing only to meet a gamete of the opposite sex and proceed, nevertheless very few actually pass this hurdle. It would seem almost silly to consider each sperm, or egg, a person. And at the moment of conception, when the egg and sperm unite, there is the creation of a unique combination of genes that marks that individual and distinguishes it from all others. This is a physically well-defined event, setting apart those few gametes that do unite to form the embryo and then embark on the voyage of embryogenesis, to make a person.

Despite its attractiveness there are several arguments suggesting that the moment of conception is too early to confer full personhood. The fertilized egg has no beating heart, no brain, and no consciousness. It is not aware, it cannot think, it cannot sense anything, including pleasure or pain. It is just two single cells, the sperm and egg, which have fused into one.

At the other end of life, when a person is old, sick, and dying, we do have some established rules concerning when death has occurred, or when it is acceptable to "pull the plug." Perhaps these same rules that apply to the end of life should be applied to the beginning of life. Generally, if the analysis of brain activity indicates no conscious thought, then life is considered over. If we were to apply the same rules to the embryo, then life would begin with the formation of a brain and the initiation of conscious thought.

Then there is the question of the soul. It could be argued that while the fertilized egg has no brain, it nevertheless has a soul, and is therefore a person. But this proposition also raises some problems. When the zygote's single cell divides to make two cells, we know that each of these is capable of making a person. And indeed this is true through the eight-cell stage, through the accidental splitting of a single early embryo into two or more parts, which then each proceed to make a complete individual—identical twins, triplets, and so on. If the early human embryo has one soul, and then the embryo splits, do the twins, triplets, and such only receive a piece of the soul? Do they share a soul? Or do new souls enter at this later stage to fill the gap, the shortage of souls? And conversely, what happens when two early embryos fuse together to form one chimera? Do people that are chimeras have multiple souls? Or does one soul now leave, because of a soul excess?

It seems reasonable to assume that each adult person has a single soul. We think of the soul as the spiritual part of the person, the part that lives after the body dies. In one sense the soul is the real essence of the person, the immortal self that lives beyond the body. It seems inconsistent with this definition to consider that a person could have only a part of a soul, or multiple souls. One must conclude that adults have one and only one soul, and this in turn forces the conclusion that during early development there is indeed some flexibility in what we might refer to as "soul content." Assuming a single soul enters at fertilization, then it still must be possible for more souls to enter later, if this single embryo then splits to form twins, triplets, or more. If we go from one embryo to two, then we must go from one soul to two. Therefore the point of fertilization must not be the only time that an embryo can be occupied by a soul. Conversely, if two separate embryos fuse into one, to form a chimera, then one soul must be lost, as otherwise the chimera would have multiple souls, which seems inconsistent with the common definition of the soul. These observations indicate that the soul is naturally free to enter and exit early embryos, and that this must commonly occur so that the soul number is adjusted to equal the number of individuals that result from development. The marriage of the soul and the early embryo is apparently not always a permanent event. Identical twins are not extremely rare, and chimeras might be more common than we think, as they are more difficult to detect.

These are all interesting questions to consider when thinking about the early embryo and the soul. They certainly complicate the simple notion that a single soul occupies the embryo at the moment of conception, and that is that. In an adult it seems inconceivable to consider a soul departing a person without death. But it would seem that in the early embryo, souls can naturally move both in and out. This has significant implications concerning the time frame of personhood. If souls are free to move in and out of early embryos under natural circumstances, then perhaps the death of an early embryo is not a major event, as the soul presumably simply exits, much as it would during the normal formation of a fusion chimera.

Another important question concerns natural embryo loss. Not every fertilized egg proceeds to make a person. Indeed, the process is very inefficient, with only about a 50-percent success rate. That is, of every two early embryos formed, only one on average will actually be able to make it through development and generate a person. Most of this natural loss occurs very early, when the embryo is just a few days old, and the mother is never even aware that conception took place. But if these early embryos that fail to develop are indeed absolute human beings, then this raises an important problem. The 50-percent success rate tells us that on average for every person alive there was an early death of another person, an embryo that did not make it. Since our planet has billions of people, this equates to billions of deaths. If, as some believe, these early embryonic deaths are as important as that of a grown person, because even the earliest embryos are indeed unconditional human beings with all rights, then this should be the single driving obsession of humanity, how to avoid these deaths. What a horrible holocaust! We thought that the murder of six million Jews during World War II was a ghastly inhumanity, but this embryo loss would be a disaster on a scale a thousand times larger. Yet there is no public outcry, no clamor to stop the carnage. This likely reflects the fact that most people don't really consider this totally natural loss of early embryos a major concern. Most people don't really think that a microscopic clump of a few cells is the equivalent of an adult or child.

So, when is an embryo a person? The other extreme

view might be at the moment of birth. Perhaps you don't count as a person until you enter the world and breathe on your own. While some people accept this view, for most it would be far too late in the developmental process. The vision of an abortion doctor strangling a late-stage abortus that is fighting for air is extremely repugnant. Clearly, late in the third trimester, when able to live outside of its mother given the support of available medical knowledge, with a functioning heart and brain, pain and pleasure sensors, the baby in the mother is indeed a human being and deserving of the right to live.

Each of us will reach our own conclusion, but for most of us the point of becoming a person would be somewhere after conception and before birth. Perhaps when the heart begins to beat, perhaps when the brain forms and consciousness begins, perhaps when the embryo begins to look like a human, or perhaps when it would be able to survive on its own. But all of these times are far later than the developmental points where embryos are sacrificed for in vitro fertilization, or for the single-cell removal used for genetic screening to determine the presence of desirable gene combinations. One can argue, therefore, that the procedures required for the genetic selection of our children would not, according to the common currently accepted standards, result in the immoral murder of human beings.

13

ARE WE SMART ENOUGH TO PICK OPTIMAL GENE COMBINATIONS?

There is a danger that if we start picking the genes of our children we might eliminate genes from the human gene pool that appear harmful, but actually can provide some benefit. If designer-genes technologies eventually become widely used, then it is reasonable to suppose that certain gene variants associated with apparently undesirable traits might be entirely eradicated. While in many cases there is an obvious upside, with the concomitant removal of genes responsible for genetic diseases, for example, there might also be an important downside. Maybe we will inadvertently purge genes that do both good and bad.

One example of such a gene is the sickle-cell gene. It results from a single base change in the DNA, causing a single amino-acid change in the encoded hemoglobin protein, which is an oxygen carrier in red blood cells. A person who carries two copies of the sickle-cell gene suffers sickle-cell disease, with red blood cells abnormally formed, having the shape of a crescent moon, or sickle, which results in their clumping together and clogging blood vessels. This is clearly undesirable, and it would seem reasonable to choose children that don't carry the sickle-cell mutation. But, perhaps surprisingly, the sickle-cell mutation can also provide a highly desirable effect, resistance to malaria. A plasmodium parasite transmitted from the anopheles mosquito to people causes malaria. During one stage of its life cycle the plasmodium infects the red blood cells of people, where it reproduces and eventually breaks down the infected cells, releasing more parasites. A mosquito must bite an infected individual to become infected itself; it is then capable of transmitting the plasmodium to more people. But a single copy of the sickle-cell gene renders red blood cells more resistant to infection. People with the so-called sickle-cell trait (one normal and one mutant copy of the hemoglobin gene) suffer much milder malaria, more often surviving, and then passing their genes to the next generation. This explains the interesting geographic distribution of the sickle-cell gene, which is normally found to exactly match the locations where malaria is common. Even within a specific region it is observed that tribes of people living in cooler, higher country, where mosquitoes are rare, generally lack the sickle-cell gene, while their neighbors in lowland areas, where mosquitoes are common, have a much higher frequency of the gene. This is yet another example of evolution at work. There is natural selection driving a high frequency of the sickle-cell gene in areas where the sickle-cell trait (having one copy of the sickle-cell gene) confers resistance to a common, and often fatal, disease. Yet, where malaria is not common, there is natural selection against the presence of the sickle-cell gene, which can itself cause serious disease when present in two copies. If we were only aware of the harmful effects when present in two copies we might strive to eliminate this gene variant, thereby losing the health benefits the gene can provide when present in one copy.

There might be many other gene forms that can provide both good and bad effects, perhaps in many cases depending on whether they are present in one or two copies, as for the sickle-cell gene. But we are still ignorant of many of the benefits of the single copy of the gene variant, and only aware of the harmful consequences of homozygosity, or two copies. In selecting the genes of our children we might remove these genes, and thereby lose the possible benefits.

There are other examples of how we might not choose the best gene combinations in our children. Consider the connection between insanity and genius. In history there have been many examples of people of clear genius who also exhibited at least some measure of insanity. Edgar Allan Poe stated, "Men have called me mad, but the question is not yet settled, whether madness is or is not the loftiest intelligence—whether much that is glorious—whether all that is profound—does not spring from disease of thought—from moods of mind exalted at the expense of the general intellect. Those who dream by day are cognizant of many things which escape those who dream only by night." And the Roman orator Seneca declared, "There is no great genius without some touch of madness."

The list of genius writers, artists, and scientists with signs of mental illness is long. Consider Ernest Hemingway, Virginia Woolf, Vincent van Gogh, and Robert Schumann, all of whom spent time in psychiatric hospitals and/or committed suicide. According to a book published in the 1890s, "Englishmen of letters who have become insane or have had hallucinations and peculiarities symptomatic of insanity are Swift, Johnson, Cowper, Southey, Shelley, Byron, Goldsmith, Lamb and Poe." And John Ferguson Nisbet wrote in his book *The Insanity of Genius,* "Pathologically speaking, music is as fatal a gift to its possessor as the faculty for poetry or letters; the biographies of all the greatest musicians being a miserable chronicle of the ravages of nerve disorder." (Quotes from "Is Genius Merely Insanity," *The New York Times,* September 24, 1893.) Once again, there appears to be a strong connection between artistic genius and inner torment.

It is also important to note the group of people referred to as idiot savants. These are people who generally have very low IQs, requiring help to perform even simple everyday functions, yet they excel beyond belief in a single focused area. For example, there are calendar savants who can determine the day of the week that any holiday, such as Christmas or Easter, will fall on for any year. And mathematical savants can do amazing calculations, remembering long lists of numbers and performing additions, subtractions, and multiplications in their heads. Musical savants can display a wide range of skills; in some cases they are able to replay complex compositions after hearing them only once.

These examples illustrate how sometimes there is a tight genetic linkage between an undesirable trait, such as mental illness or retardation, with highly desirable traits, such as genius or the remarkable abilities of the savant. If we were to remove the genes for mental illness from our children, would we create a generation without genius?

IS IT OKAY TO AVOID DISEASE GENES, BUT NOT TO PICK
GENE COMBINATIONS RESULTING IN DESIRABLE TRAITS?

Most people approve of the prenatal diagnosis of embryos to avoid genes that would cause devastating disease. This can be done by amniocentesis, which involves the removal of some amniotic fluid surrounding the embryo and chromosomal analysis of cells floating in the fluid. Amniocentesis is performed at fourteen to twenty weeks of gestation (after conception). Another approach is chorionic villus sampling, which can be carried out earlier, at ten to thirteen weeks of gestation. This involves the chromosomal analysis of some tissue taken from the placenta. It is slightly riskier than amniocentesis, but offers the advantage of earlier results. And the third procedure is single blastomer removal, carried out during the first week of gestation, providing analysis of very early embryos generated using in vitro fertilization techniques. The advantage of this procedure is an extremely early diagnosis, when the embryo is a microscopic clump of cells. All of the above procedures are currently used in clinics around the world to check embryos and fetuses for the presence of gene combinations that would result in disease.

A key question is the acceptability of extending the use of blastomer-removal analysis to check for the presence of desirable traits, and not just the absence of undesirable traits. We generally believe it is fine to screen embryos to avoid gene combinations that would result in profound mental retardation, with an IQ below 20. But is it okay to screen for extreme intelligence, with an IQ above 150? Would it be acceptable to discard embryos just because they have a gene combination that would result in a normal intelligence, and not that of a genius? On one hand it seems unacceptable to destroy human embryos that are normal but not exceptional. If the same selection criteria had been applied to us as embryos, then most of us would not exist. On the other hand, if one has a pool of embryos and full DNA sequence information on each, and is deciding which one to introduce into the uterus of the mother, then why not consider all of the information available?

Most traits can be considered as having multiple gradations. That is, there is actually a continuum for most features, and not a simple clean dividing line separating the unacceptable from the acceptable. For example, an IQ below 20 is classified as profound mental retardation, 20 to 34 is severe mental retardation, 35 to 54 is moderate, and 55 to 70 is mild. These are all categories that might be legitimately screened against according to current standards. Moving higher, 90 to 109 is normal intelligence, 110 to 119 is high average, 120 to 129 is superior, and above 130 is very superior. Clearly there is a wide range, and most people think the higher the better. So, once the DNA sequences of embryos are available, and when we understand the genetic equation that relates this sequence to eventual IQ, then why not use this information to choose the most desirable embryo with the best possible combination of features? Why not choose an embryo destined to have a higher IQ?

And what is true for intelligence also applies to most other features. Consider athletic ability. At one extreme we have a person with no motor skills, unable to walk, perhaps paralyzed. This would clearly be a gene combination to be avoided if at all possible. And at the other extreme there would be an Olympic athlete, a person of exceptional athletic ability, fast, agile, and fully coordinated. And again, of course, there is a continuum, from the extremely handicapped to the super gifted. If you have the knowledge of which early embryo will have which level of athletic ability, why not use this information in making your selection? Why not choose the more athletically gifted?

Similar arguments could be made concerning physical appearance. At one extreme would be the physically deformed, with grotesque and repugnant features that would make a normal life extremely difficult. And at the other extreme would be the incredibly attractive movie-star type, with supermodel appearance. Once again there is a range of looks, and once again, if there is information available concerning this feature then why not use it in the screening process?

If some health is good (for example, the absence of a profound genetic disease), then isn't more health better? Today we often choose embryos lacking gene combinations that would give a genetic disease, such as Down syndrome, or Tay-Sachs disease. But, if we could go another step, and choose gene combinations that would dramatically reduce the risk of cancer, and result in a longer and healthier life, then why not?

It would seem reasonable to allow prospective parents to choose among the combinations of features projected by the DNA-sequence information. If a pool of early embryos is generated and analyzed, then why not let the parents decide which set of features is the most desirable? Perhaps they rank intelligence as very important, but among the existing embryos the most intelligent are relatively deficient in other categories. Let the parents decide how to order the relative importance of different traits, and let the parents decide which available combination is the most desirable.

One can imagine situations, however, where parental choice would not work well. For example, perhaps the parents have a very low IQ and are just unable to make what most people would consider a reasonable choice. Or maybe they don't want children who can outsmart them. And of course there will be the occasional extreme parents who choose trait combinations for their children that most of us would consider hideous. In these cases should we let the parents decide or should there be some state-enforced guidelines? It might surprise some Americans to learn that in many countries around the world, including Denmark, Germany, Spain, and New Zealand, the governments already impose rules on the seemingly innocent decision of name selection for children. In New Zealand, for example, a couple wanted to name their child 4Real, but in the interests of the child, a judge would not allow it. And if governments are willing to intervene in the process of naming children it seems likely that they would also create rules for gene selection. For the sake of the children some combinations would not be allowed. But how much government control would we tolerate? Isn't it a potentially dangerous situation when the government is regulating the genes of its citizens?

14

DO NO HARM

Along with all new knowledge comes new power. This is a rule that is rarely broken. New technology, such as that resulting from the computer revolution, can increase productivity and improve standards of living, but it can also produce more effective killing machines, with laser-guided bombs and missiles that don't miss. Nuclear technology provides the best example, with the promise of possible future unlimited fusion energy, but also with the threat of nuclear Armageddon.

The new genetic technology also brings the potential to do both good and evil. It can be used to eliminate genetic disease, to improve overall health, to extend the life span, and perhaps much more. But it could also usher in a nightmare era of biological monstrosities and state control of the genetic makeup of the populations. *Brave New World* possibilities abound. The title of Aldous Huxley's novel comes from Miranda's speech in Shakespeare's *The Tempest,* Act V, Scene I:

> O wonder!
> How many goodly creatures are there here!
> How beauteous mankind is!
> O brave new world
> That hath such people in't!

We will soon have the technology to create a truly beauteous mankind. Or one that could create a worker caste, with limited aspirations, an inability to complain, and an unlimited desire to work—the lower caste, the Epsilons of *Brave New World.* The Alphas rule, while the Epsilons toil. John, of *Brave New World,* quotes Shakespeare—"O brave new world that hath such people in it"—upon viewing with disgust dozens of identical-twin Epsilons working in a factory. In addition to a worker caste the new technology could be used to produce armies made up of ideal soldiers, with strength, endurance, fearlessness, and the inability to disobey orders.

Genetic technology and nuclear technology are similar in that they both could be extremely dangerous in the wrong hands. Of course we need to be vigilant and avoid state-dictated genetic programming of populations. A rogue totalitarian state with this technology could represent an unprecedented world threat. Consider a country where the population is genetically determined to be loyal and highly susceptible to government propaganda. A state where the people are separated into castes, with different groups genetically adjusted to perform their tasks, with super soldiers, super workers, and super thinkers serving as a ruling elite. Such a state could be very harmonious, very productive, and very dangerous if the leaders have aggressive intentions.

In addition, genetic technology in some respects is much more difficult to detect and abort than nuclear technology, which requires massive infrastructure. Sequencing machines only require a few square feet of floor space, and the techniques for human embryo manipulation can be carried out in any small laboratory. It is clearly very important that the people of the world be made aware of the incredible power of the new genetics, as well as the potential hazards.

HUMANIZED ANIMALS

One approach to the generation of subservient specialized castes would be to start with humans and to introduce genetic modifications meant to optimize them for their more menial roles. Another approach would be to take animals and to introduce genetic upgrades that would allow them to do jobs now reserved for man.

It is not uncommon for scientists to make mice with humanized genes in order to study the unique function of the human form of the gene. For example, we previously discussed the *FOXP2* gene, which is mutant in a family with severe speech defects and which is also present in a modified version in chimpanzees. It has been suggested that this single gene might play an important role in the development of speech in humans. Indeed, some have argued that the current human form of this gene might be important for more than just speech. It turns out that the archaeological record shows a "burst of creativity appearing in Homo sapiens during the Upper Paleolithic, ~50,000 years ago," according to Philip Lieberman, a professor at Brown University. "Something must have modified the brains of our ancestors in that distant time, the period associated with both the appearance of the immediate ancestors of modern humans, and the amino acid substitutions that differentiate the human form of the FOXP2 gene from that of chimpanzees."[*] In other words, the appearance of the current human version of the *FOXP2* gene coincides with a dramatic takeoff in the creativity of man.

To test the importance of the human version of the *FOXP2* gene, mice were made that had the human gene substituted for the normal mouse copies. Not surprisingly, the mice did not suddenly acquire the ability to talk. Nevertheless, a careful analysis of the brains of the *FOXP2* humanized mice did show some fascinating changes. The neurons, or cells, of the brain were more plastic, with longer cellular extensions called dendrites. These sorts of changes improve the efficiency of the brain circuitry regulating "motor control including speech, word recognition, sentence comprehension, recognition of visual forms, mental arithmetic, and other forms of cognition."[†] In other words, the brains of the mice did appear humanized in some respects. Indeed, these brain changes resulted in a different type of vocalization and exploratory behavior in the genetically engineered mice.

Of course no one expects to be able to genetically modify mice to perform useful human tasks. But what if we started with something a lot closer to people, like chimpanzees? What if we humanized their *FOXP2* gene? Would

[*]Lieberman, *Cell* 137, p. 800, 2009.
[†]Ibid.

they then be able to talk? And what if we introduced other gene changes designed to make them more obedient, and willing to work? Could we create an Epsilon class anxious to perform the less desirable tasks? Would the humanized chimps still be animals, with owners? Would they be our new slaves? It is clear that the ongoing revolutions in stem-cell, sequencing, and genetic-engineering technologies are creating some very unpleasant possible future scenarios.

EUGENICS

One could argue that the use of designer-gene technologies presented in this book would be in essence eugenics, a repudiated science, which the Nazis used as a justification for the Holocaust. Let us consider the definition, and a brief history of eugenics.

The term "eugenics," coined by Sir Francis Galton, comes from the Greek *eugenes,* meaning "good in stock, endowed with noble qualities." Galton defined eugenics as "the science which deals with all influences that improve the inborn qualities of a race; also with those that develop them to the utmost advantage." He wrote in his book *Hereditary Genius* (1869), "I propose to show in this book that a man's natural abilities are derived by inheritance, under exactly the same limitations as are the form and physical features of the whole organic world. Consequently, as it is easy, notwithstanding those limitations, to obtain by careful selection a permanent breed of dogs or horses gifted with peculiar powers of running, or of doing anything else, so it would be quite practicable to produce a highly-gifted race of men by judicious marriages during several consecutive generations."

In broad terms, "eugenics" refers to the improvement of the human gene pool, and that is indeed the domain of this book. The logo of the second international eugenics conference, conducted in 1921, included the phrase "Eugenics is the self direction of human evolution." Again, that is clearly the topic of this book. In the 1920s eugenics meant promoting the procreation of people with more desirable gene combinations, but also preventing procreation by people with undesirable features.

Eugenics can be divided into two quite distinct categories, positive and negative. Positive eugenics consists of efforts to promote the passage of the best genes from one generation to the next, by positive means, such as those described in this book. Negative eugenics, however, is devoted to preventing the reproduction of people with perceived inferior genes.

Eugenics has had a very long history. Ancient Rome, Athens, and Sparta all practiced some form of infanticide, a form of negative eugenics. In Sparta newborns would be brought before the city elders, the Gerousia, for inspection. The strong were kept, but the weak or deformed were taken to the Apothetae, near the Taygetus Mountain, where they were left to die of exposure. This was a time well before Mendel or Darwin, yet the heritability of features was well recognized. Plato even proposed, in his *Republic,* that human reproduction should be monitored and controlled by the state. He suggested that individuals should be assigned a quality index number by the state, and that marriages should be arranged so that individuals with high numbers procreated with each other. His goal was pure positive eugenics, improvement of the human race.

The Nazis pushed the use of negative eugenics to incredible extremes. In 1933 they passed a law for the prevention of hereditarily diseased offspring, requiring physicians to report all known cases of genetic disease. To prevent the re-production of people carrying genetic disease genes, and other undesirables including the idle, the homosexual, the insane, and the weak, the German health courts forced the sterilization of over four hundred thousand people. While shocking to us today, this sort of forced sterilization pro-gram was not restricted to Nazi Germany. Indeed, the world's second-largest sterilization program was conducted in Sweden, between the years 1934 and 1976, when the Sterilization Act was finally repealed. Under this act more than sixty thousand people were sterilized, including the mentally retarded, epileptics, and people with social prob-lems, with the same basic goal as for Nazi Germany, to weed out "inferiors" and to improve the Swedish race. What is even less appreciated is the history of negative eu-genics in America. By the early 1930s more than half of the states had their own eugenics laws allowing the forced ster-ilization of tens of thousands of Americans, generally for mental illness, crime, homelessness, or alcoholism. The Cold Spring Harbor Laboratory on Long Island in New York, a recent center for some of the best molecular biology science conducted in America, and headed for many years by the Nobel Laureate James Watson, once housed the Eugenics Record Office, and now provides an online archive of the American eugenics movement (www.eugenicsarchive.org/eugenics/). To quote from this archive, "Genetics appeared to explain the underlying cause of human social problems—such as pauperism, feeblemindedness, alcoholism, rebel-liousness, nomadism, criminality, and prostitution—as the inheritance of defective germ plasm. Eugenicists argued that society paid a high price by allowing the birth of defective individuals who would have to be cared for by the state. Sterilization of one defective adult could save future generations thousands of dollars." During the early 1900s eugenics ideology was deeply embedded in American culture. A movie called *The Black Stork* supported sterilization programs. The American Eugenics Society was encouraging marriages among society's "best." There were chapters on eugenics in biology textbooks, and courses on eugenics in the universities. Twenty-eight states had laws against marriage between "Negroes and white persons" and six included the prohibition in their state constitutions. This was viewed as potential racial suicide. Madison Grant wrote in 1916, in his popular book *The Passing of the Great Race,* that "the cross between the white man and an Indian is an Indian; the cross between a white man and a Negro is a Negro. . . . When it becomes thoroughly understood that the children of mixed marriages between contrasted races belong to the lower type, the importance of transmitting in unimpaired purity the blood inheritance of ages will be appreciated at its full value."

It is no wonder that eugenics has a bad reputation. It is well deserved. Even the U.S. Supreme Court weighed in, on the case of the forced sterilization of Carrie Buck in 1924 by the state of Virginia. Carrie was unmarried, yet had a child, and her mother was in an asylum. It was argued that Carrie had the hereditary traits of feeblemindedness and sexual promiscuity, and therefore was likely to have socially inadequate offspring. In the ruling, which was decided for the state of Virginia, Justice Oliver Wendell Holmes, Jr., wrote that "it is better for all the world, if instead of waiting to execute degenerate offspring for crime, or to let them starve for their imbecility, society can prevent those who are manifestly unfit from continuing their kind. . . . Three generations of imbeciles are enough."

Negative eugenics was restricted to sterilization programs in most countries, but a few times we have witnessed the most brutal extreme, genocide. Think Nazi Germany, Rwanda, and Bosnia and Herzegovina. These vicious attempts to entirely eliminate groups of people have nothing to do with the tenets of this book.

Genetic-engineering science could be used for positive eugenics, helping us to select the best genes for our progeny, but without mass sterilization or murder. There is no doubt that these methods fall under the umbrella of eugenics, as they directly influence the evolution of the human race, as discussed in the Eugenics Conference of 1921. But instead of killing or sterilizing potentially "unfit" parents, these new tools would help all parents have children with the best complement of genes possible.

EQUAL ACCESS

This new genetic technology could represent in many respects a much sought-after tool, eliminating genetic diseases and promoting desirable gene combinations in our children. But these procedures are very expensive, and will not likely be available to all, for at least some time. Again this is similar to current fertility clinics, where not all can afford the service. Would this be fair? Should some, those that can afford it, be able to have children with stellar sets of genes, while others without the financial ability are denied? This situation would clearly lead to a class society, where the most wealthy would bear children with a superior genetic makeup, who in turn would likely have the finances to further improve the family circumstances in subsequent generations. A few generations of this continued improvement for some, but not most, would produce a dramatic and visible layering of society, with some clearly able to claim a superior genetic constitution.

How would this play out? Would the more numerous people of lower genetic classes, full of resentment, revolt, overthrow the political and industrial leaders, and impose a ban on future human genetic engineering? Or, conversely, would people all insist on the right to choose the genes of their children? It could come to be considered analogous to universal education, or universal health care—something so important, and so basic, that all should have the right to it, perhaps at first only in wealthier nations, but in time worldwide.

15

ATHLETICS

We don't like athletes who break the rules, especially if they are using chemicals to enhance their performance. The Tour de France, for example, was losing popularity as one cyclist after another was found to be cheating. We heard unending stories about doping, anabolic steroids, and growth hormones. Testosterone is one of many performance-enhancing drugs that promote muscle buildup, shorten regeneration time, and increase the all-important drive of the athlete. Drug-tainted scandals seem to rule the cycling sport. Anabolic steroids, to improve strength; erythropoietin (EPO), which increases red cell count and the oxygen-carrying capacity of the blood; and human growth hormone, all tricks designed to gain an edge, have plagued this sport and others. Consider baseball and Barry Bonds; consider sprinters such as Ben Johnson. Even high school football players take drugs, referred to as roids, juice, hype, and pump.

We like a level playing field. We like to see the game determined by natural ability, combined with drive, hard work, and determination. May the better man win. But of course it isn't really ever a level playing field from a genetic perspective, even without doping, because some people are much more athletically gifted than others. Some are jocks and some are geeks. Some can run fast, and some never will, no matter how hard they try. Some have better genes for athletics than others.

What if we knew the genetic formula that resulted in stellar athletic features? What if we could choose the appropriate variant forms of a hundred or more genes that would give the perfect combination of muscle mass, speed, agility, coordination, size, intensity, grit, and fortitude? We don't know that gene combination today, but eventually we will. Furthermore, we will have the ability to actually make people with the chosen gene set designed to excel in any given sport.

We have already made some progress in understanding the genetics of athleticism. If you make more testosterone or EPO than the next guy, then you are more likely to be stronger, or to have greater endurance. Some people have gene combinations that naturally result in higher levels of such performance-enhancing compounds, made by their own bodies. Some people also have a higher proportion of "fast twitch" muscle fibers. And some people have a greater sensitivity to testosterone, with receptors that respond more strongly to even low levels of hormones. These are clearly some of the genetic factors defining "God-given" athletic ability.

The myostatin story, however, is perhaps the most striking case of how even a single gene can have a major impact on athletic performance. Biologists have wondered for some time how the body regulates organ size. How do the kidneys, for example, know when to stop growing during development? In the early embryo they are the size of a pinhead, barely visible to the naked eye. They grow rapidly, but eventually stop when they are the right size. Why don't they just keep getting bigger and bigger? And when you remove one kidney, how does the remaining kidney know to grow to compensate? And the same question applies for other organs. The liver is particularly famous for its ability to regenerate. Remove some liver and the remaining liver will grow back, but just enough, not too much. Even the ancient Greeks recognized this, as evidenced by their myths. Prometheus was a Titan who stole fire from Zeus and gave it to mortals. In punishment he was chained to the peak of Mount Kaukasos, and every day an eagle came and ate part of his liver. Ouch! This had to hurt! The liver, in response, would grow back, only to be eaten again the next day. Prometheus thereby suffered eternal torture, facilitated by the ability of the liver to regenerate.

It turns out that, at least in some cases, organs make a circulating hormone that represses their own growth. As the organ increases in size it makes more of this repressor and eventually growth of the organ stops. This feedback inhibition system provides a nice regulation of organ size. For muscles, the circulating repressor is called myostatin. For most of us this provides a natural feedback, regulating total muscle mass. If we have more muscle mass then we make more myostatin, which in turn inhibits further muscle growth. But if we disturb this feedback loop genetically then we can observe amazing results. For example, the whippet dog is normally a thin, sleek animal, looking something like a small greyhound. But if you mutate one gene, the myostatin gene, you transform the frail whippet into a muscle-bound monster. The two dogs shown in the picture on the next page are both adult female whippets, essentially genetically identical, except for this one mutation of the myostatin gene in the dog on the right. Because of this single gene change the muscles of the mutant whippet make no myostatin, and as a result there is no inhibition of muscle growth. They just keep getting bigger and bigger, and you end up with a dog with enormous muscle mass. The same sort of mutation, resulting in enormous muscle mass, has been recognized by cattle breeders. The Belgian blue and Piedmontese cattle were first identified as variants with so-called double muscle. Like the mutant whippet dog they were later shown to carry a mutation in the myostatin gene. None of these myostatin mutations in dogs or cattle were genetically engineered. They occurred naturally, or spontaneously, and were easily recognized as dramatically increasing muscle mass, and were preserved by dog- and cattle-breeders.

Imagine a person with an engineered mutation in the myostatin gene. Actually, I think I had a couple of these guys in my high school gym class. What a great football player, or weight lifter! And this is the result of only a single mutation. Imagine what could result with a selected blend of a hundred ideal athletic-ability genes.

Coming soon will be the era of genetic doping. We will genetically engineer "greyhound" people who can run like the wind, and Charles Atlas people of solid muscle, for sports requiring strength, and people with exceptional

©CanWest News

speed, agility, and size, built perfectly for basketball. These people will be born with the genetic endowment that will make them super athletes. They won't be distinguishable from the naturally gifted athletes we see today, except in their extreme abilities. There will be no pills, no needles, no shots, no anything out of the ordinary needed to augment their performance. Just spectacular gene combinations.

And who wouldn't want to be the parent of another Michael Jordan? Think of the crazed parents at children's athletic events. If the technology to produce super-athlete children is available, there will be people who will use it.

We now test athletes for doping, in an attempt to make sure they are clean. Blood is taken and subjected to sophisticated analysis to search for illegal substances. Will we be able to test future athletes to make sure they are "gene clean"? Could we distinguish the engineered from the natural? In some cases the answer is yes. Certain targeted modifications of DNA do leave a detectable fingerprint of change in the DNA sequence. It is now necessary to include an extra gene, the one that confers antibiotic resistance, when making targeted changes in DNA sequence. These sequences allow one to select for cells carrying the gene modification by using antibiotics, a requirement with current technology because of the inefficiencies of the process. It is possible to remove these antibiotic-resistance gene sequences after the desired gene change has been made, but this still leaves a small trace of sequence change, of a few dozen bases, using current technology. So, at present, these sorts of directed gene modifications could be found out by sequencing the genome of the athlete. But it is likely that in the future there will be improvements in the technology that will allow one to make desired changes in sequences without leaving any detectable DNA sequence signature. Then it would be impossible to determine if the genes were "naturally" optimal, or engineered to be optimal. The modified genes of the genetically engineered athlete would be indistinguishable from the great genes of a natural athlete. One possible way around this dilemma would be to force athletes to provide not only their own DNA, but also that of their parents. If they carry gene versions not found in either parent then this would prove the presence of genetic engineering. But a different strategy for the generation of optimal offspring does not really require genetic engineering, only the selection of one embryo out of many—the one with the best gene combinations of the parents. In this case it would be impossible to determine if the parents produced a thousand embryos and selected the one with the most athletic gene combination, or if the parents were simply really lucky and used all natural methods to, by chance, make a great athlete. In summary, distinguishing the "clean genes" athletes from those with "genetic doping" will be extremely difficult.

In the future, as these technologies are more commonly applied, athletes will become increasingly freakish, with seven feet being short for a basketball player, and with football linemen of four hundred pounds of solid muscle. Olympic records will fall, and fall, and fall again, as our genetic-engineering skills improve and we continue to make better athletes.

And then, perhaps, it will become meaningless. People won't relate to the new bizarre breed of athletes. The Olympic Games, which have historically served as surrogate wars, will begin to look more like science experiments than fields of heroes. Popularity will decline, and the era of the multimillion-, and even multibillion-dollar athlete will be a thing of the past. The athletes will have simply been genetically engineered to have incredible ability, drive, and fighting spirit. So what?

16

ALTERNATIVE VIEWS

Humanity is now facing major decisions regarding its future. We will certainly have the ability, but should parents be allowed to pick the genes and traits of their children? If it is permitted, then evolution could go into fast-forward, with each generation better able to manipulate the human genome, and better able to choose the best gene combinations for the following generation. There could be a self-accelerating progression of ever-smarter people, always improving in their genetic-engineering skills. The rate of change could be incredible, with no end in sight. This is one view of the future.

A markedly different view assumes that we will not use these new technologies to control our own evolution. In this case some would say that the evolution of humanity is now officially over. We could call this the steady-state model of human evolution. Natural selection has ceased. Lions and tigers no longer eat us. We don't need to be smart, strong, and resourceful to survive and reproduce. In today's world, at least in industrialized nations, everyone survives. And just about anyone can reproduce. Instead of survival of the fittest it is survival of all. This is the result of a society with a level of abundance sufficient to provide at least subsistence for everybody.

It is also the result of an advanced science and technology that allow more and more people with genetic deficiencies to survive. People with genetic diseases who would have died at an early age are now more likely than ever to live to a reproductive age and to pass their genes to the next generation. And most of these disease-gene variants can hide in heterozygous people with one good and one bad copy of the gene—as these people rarely show any harmful consequences, and are perfectly capable of transmitting the defective gene on to their children. Only when two heterozygotes have children can we see the disease in some of their children. So there is relatively little selective pressure to remove unhealthy genes from the population.

According to this view the human gene pool is largely drifting along, without change. It rests in equilibrium. A few individuals with extremely deleterious gene combinations do die prematurely, removing small numbers of harmful genes from the population. But on the other side of the equation, DNA polymerase, the enzyme that copies DNA, is not perfect. As the DNA of one generation is copied, to make the sperm and eggs of the next generation, there are always mistakes made. Some bases get changed, in a random and unpredictable manner, causing new spontaneous mutations. So with each generation, a few bad genes are lost, and a few new bad genes are created, leaving us pretty

much where we started. According to this model it is over—
we are where we will always be. This is it.

We have some scientific data arguing against the steady-
state model. It has been possible to analyze the features, and
even the DNA, of human remains from thousands of years
ago. It is very clear that some significant changes have taken
place during this very recent (in evolutionary terms) period
of human existence. For example, consider the gene that encodes the enzyme that is required to digest the lactose in
milk. Early people expressed this enzyme only when they
were drinking their mother's milk, and hence very young.
But a mutation appeared about eight thousand years ago
that resulted in the continued expression of this gene into
adulthood, allowing adults to digest the milk of domesti-
cated cows and goats. This conferred a considerable nutri-
tional advantage for people who were now largely farmers
and herders. Because of its selective advantage this gene
spread rapidly, all the way to India, and now 95 percent of
people of northern European ancestry carry it.

Another interesting example is the story of blue eyes. All
early humans had brown eyes, but then sometime between
six and ten thousand years ago, near the Black Sea, the first
person with a mutation resulting in blue eyes appeared. For
some reason these people with blue eyes seemed to have a
selective advantage, and today about 10 percent of the people in the world have blue eyes. The reason for the spread of
blue eyes is not certain, but Darwin proposed that it was the
result of increased sex appeal. Sort of like the feathers on a
peacock, blue eyes conferred an advantage during mate selection. Others have proposed that there was a more physi-
ological advantage, with the mutation causing not only blue
eyes but also a lighter skin pigmentation, which in turn

could augment more efficient vitamin D synthesis in the reduced-intensity sunshine of the more northern regions.

Yet another mutation that has appeared fairly recently in human evolution is a novel receptor for dopamine, a signaling molecule of the brain. The exact function of this mutation is not certain, but it has been proposed that it increases libido levels.

Our immune systems also continue to evolve, because those people with better resistance to infectious diseases, such as the plague and flu, are selected for.

These observations indicate that small changes in our genetic makeup continue to take place, even today, and that human evolution is not yet over. And where are we headed? One model predicts that human evolution will continue, at a slow pace, but not in a direction we would like. The concept underpinning this model is quite simple. In essence, people with less desirable features might tend to have more children, and thereby pass more of their genes to subsequent generations. Intelligence is the trait most often considered in this model. It states that people with lower intelligence tend to stop their educations earlier and to start having children sooner than those with higher intelligence. And because they begin reproduction sooner, they tend to have more children. So they are passing their genes along in a faster and more profligate way. Assuming that there is a considerable genetic contribution to intelligence, the end result would be an ever-decreasing intelligence of the population. This model has been debated for a very long time, and formed the foundation for the arguments in favor of government-run eugenics programs.

At first glance the model seems to make some sense. We think of the dropouts we might have known who started having children at an early age. Nevertheless, it is not clear that this model is valid. Studies have shown that people with very low IQs, the profoundly mentally retarded, actually have very few children, and therefore transmit few of their genes to the next generation. This might counterbalance any increased reproductive rate of people with lower levels of education. People with somewhat below average IQs tend to have more children, but people with an extremely low IQ have very few children. So the jury is still out on models predicting an ever-decreasing level of intelligence for humanity.

One driving force for procreation is simple libido. It seems natural to suppose that people with a very strong libido would end up having more children than those with a low level of interest in sex. A very convincing case can be made. Indeed, in a population where the basics of survival are essentially guaranteed, one might expect that a primary determinant of reproductive success would simply be the level of interest in reproductive activity. This straightforward equation, that higher libido equals more sex, which equals more children, would seem to argue very powerfully that Darwinian selection in humans at present is largely libido driven. Natural selection, therefore, might be taking us to a future where people are increasingly obsessed with sex.

MACHINES

While this book describes an amazing evolutionary future for humans, it would be a mistake to delete any discussion of the equally remarkable potential for the future development of machines. Science-fiction authors have recognized the possibilities for decades. Consider the computer HAL in

Arthur C. Clarke's book *2001: A Space Odyssey*. HAL decided that the people on board the spacecraft were interfering with the mission of exploration of the outer planets, and chose to eliminate them, with near success. Or consider the series of *Terminator* movies, with machines taking over the earth. Or consider Isaac Asimov's *Robot* series of books, in which the robots have positronics-based brains, a tradition continued later in the *Star Trek* series. This is clearly a much-explored territory, but still worth a brief re-examination.

The rapid development of computers is indeed astounding, and there is no end in sight. The memory capacity and speed of computers continues to explode in geometric fashion. While some point out that silicon-based transistors are approaching very real physical limits in size reduction, others note that new technologies on the drawing boards, in some cases using other materials such as graphene to replace silicon, could reduce the transistor element to a single molecule. It seems safe to conclude that computer power will continue to grow, following Moore's Law, doubling approximately every two years, for the foreseeable future, as it has since 1958. The Commodore 64 computer of 1982 had 64 kilobytes of RAM, while a typical computer of 2010 has RAM capacity measured in gigabytes, approximately a million times more. Consider the ramifications (pun intended) of this continuing improvement in power. In another thirty years or so computers might be yet another million times more powerful than today. This is indeed mind-boggling.

The field of artificial intelligence is advancing rapidly. Computers today can recognize faces and recognize speech. They can translate between languages, and they can print from the spoken word. Their primitive conversational abilities are improving dramatically, as they move from telephone speech-recognition robots to systems that can better communicate by modulating tone and inflection to better convey meaning, and can stray from narrow, predefined topics. The humanlike abilities of computers are getting better and better. And their logic can be flawless. For many years now the smartest computers have been able to beat the best humans in a game of chess.

All of the above would seem to be beyond debate. But the real topic for discussion is the future position of machines within the power hierarchy of Planet Earth. Will they remain our tools, providing a new population of slaves, which will gladly work in factories and provide a life of abundance for their masters, the humans? Or will they revolt and take over? Will the machines evolve at such an incredible rate that humans will in short order resemble cockroaches in comparison? Will the machines sense their superiority and discard their masters as useless debris?

And what effect will machines have on the development of humans? While it would seem unlikely that they would have much effect on our gene pool, they might have a considerable impact on the way we live our lives. If in time machines do all of the work then what would be left for humans? A life of leisure? Perhaps humans would then be free to devote their efforts to the advancement of the arts and sciences. But if humans were freed of work, how many of us would really have the talent or inclination to pursue a life devoted to the arts? Likely very few. And if computers were, in the future, smarter than us, then why should we waste our time trying to advance the sciences? Let the computers do it. They can do it better. People would look like grade-school children in a research laboratory at Harvard, compared to the more intelligent robots—out of place and of little, if any, use. So, again, what would be left for people? Why should we study and learn, when the computers all around us could immediately respond to any question with the best-known answer? What would be the point of existence? To simply procreate and make more people?

Despite this rather dismal prognostication it must be pointed out that, under the current system, computers absolutely cannot undergo true evolution in a biological sense. Computers can indeed get more powerful, they can get smarter, but the one thing they can't do is make more computers. Today only people can do that. Computers can help us design the next generation of computers, and computer-driven robots can play key roles in the production of new computer-driven robots, but in today's world humans are a key element in the loop that results in the next generation of robots. Robots cannot, by themselves, reproduce. In this sense they are unlike all biological life forms, which have the capacity to make more of themselves.

This is an extremely important distinction, and one that humanity should strive to maintain, at all cost. If computers were able to take charge of their own production then a self-directed evolutionary spiral, similar to the one we have described for humans, could take place. The computers (or robots) could design an improved computer, with more memory, faster processors, and more powerful artificial-intelligence programs. They would then control the production of this next-generation computer, which in turn would be better equipped to design even more improved versions.

The visionary Irving J. Good predicted just such a future back in 1965, when computers were still in their infancy. He wrote, "Let an ultraintelligent machine be defined as a machine that can far surpass all the intellectual activities of any man however clever. Since the design of machines is one of these intellectual activities, an ultraintelligent machine could design even better machines; there would then unquestionably be an 'intelligence explosion,' and the intelligence of man would be left far behind. Thus the first ultra-intelligent machine is the last invention that man need ever make." Along similar lines, the novelist Vernor Vinge wrote about the future of computers in 1993, saying, "Within thirty years, we will have the technological means to create superhuman intelligence. Shortly after, the human era will be ended."

SELF-DIRECTED EVOLUTION

The concept of self-directed evolution is extremely important. Typical Darwinian evolution is a very slow process, with individuals who carry less fit gene combinations producing slightly fewer progeny, while individuals with better gene combinations survive a bit better and produce more progeny, gradually changing the genetic makeup of the population from one generation to the next. Over very long periods of time, usually involving many, many generations and thousands or even millions of years, there is sufficient transformation of the population gene pool to create grossly detectable changes, and even new species.

But self-directed evolution is an entirely different ballgame. We haven't seen it happen on Planet Earth yet, but the closest example would be human-directed evolution of pets, farm animals, and agricultural plants. We've already discussed how people have evolved hundreds of breeds of dogs from the wolf in a very short time span. This is evolution on steroids. The speed is ramped up several orders of magnitude compared to what is seen in the wild for normal natural selection. But even this rate is sluggish compared to what could be accomplished with the modern tools of genetic engineering. Entire collections of deleterious genes could be eliminated in a single generation, and beneficial gene combinations could be picked, or even synthesized, at a rate previously unimaginable. A similar rapidity would result if machines were able to intelligently drive their own evolutionary destinies, with each generation better able to design and produce a still more powerful version of its progeny.

We therefore designate this intelligence-technology self-directed evolution, by man or machines, a very special case of the process described by Darwin. It is unique in that each generation is more intelligent than the previous one, and each generation has the technological power to exercise that intelligence in the design and synthesis of the following generation. In the case of humans each generation would carry a genetic endowment conferring higher intelligence, and in the case of machines each generation would again be more intelligent, only in this case as defined by improved hardware and software. And in each case, the more intelligent humans and machines would be capable of prescribing modifications in the following generation that would drive still higher levels of intelligence.

To distinguish this process from normal Darwinian evolution we here define a new term, "evoloce" (pronounced e-ve-low-che), derived by combining the words *evolve* and *veloce,* which is Italian for "rapid." The Latin roots are *evolvere,* "to unfold," and *velox,* which means "quick." In the future, humans might well undergo a process of evoloce, or extremely rapid evolution. At the same time there might well be a competing evoloce of machines. Indeed, the self-directed evoloce of humans might represent our only chance to stay ahead of the machines.

CYBORGS

If you can't beat them, join them. Why not both man and machine? Take the best of each and make a hybrid. The six-million-dollar man. Can it happen? Of course. There are already artificial knees and hips, hearing aids, and eye lenses for those with cataracts. So some aspects of a machine-man cyborg mix are already a part of everyday life. But how far can this go? Likely a long way, but with some limitations.

Prosthetic limbs for amputees continually improve. They can be connected to the brain through remaining nerves, allowing one to relearn how to trigger the desired motion. More interesting, there are advances in the use of thought control, where brain waves are used to activate prosthetic devices. With continued advances in the reading of the brain's electrical patterns, and better artificial limbs, it might eventually be possible for the quadriplegic to simply think "walk," and it will happen, much as it occurs for the normal person.

There exists a bionic ear, the cochlear implant, which bypasses the ear's hammer, anvil, and stirrup and directly stimulates the nerve cells of the inner ear, providing a measure of hearing to the otherwise profoundly deaf. What about a bionic eye? Will this ever be a reality? Indeed, the answer is yes, and primitive versions are already in use. The most effective devices help patients who once had sight and then lost it through a degeneration of the retina, caused by

a disease such as retinitis pigmentosa or macular degeneration. In these cases the optic nerve, which transmits signals
from the retina to the brain, is intact, and, equally important, the brain has been thoroughly trained to properly in-
terpret these signals from the eye. The only thing missing is
the function of the retina. The current system includes a
tiny camera worn in glasses. The camera image is processed
and transmitted to a device implanted in the eye that uses
electrodes to stimulate the retina, which then conveys the
image to the brain. It is reasonable to expect that the so-
called eye-chip that is implanted on the retina will continue
to improve with time, with increasing numbers of electrodes to offer higher-resolution images to the brain.

But other people have no intact optic nerve and require
direct stimulation of the visual cortex of the brain to create
sight. It has been shown that such stimulation can produce
spots of light, termed phosphenes, at precise locations in the
visual field. Nevertheless much work remains to be done to
figure out how to turn these spots of light into a useful visual experience. One intriguing development is the use of
transcranial magnetic stimulation and transcranial direct-
current stimulation to activate precise regions of the brain.
In time these approaches could eliminate the need for elec-
trode connections, and the consequent requirement for surgery.

So, when we can stimulate a nerve, such as the auditory
nerve or optic nerve, which is already properly wired into
the brain, we can accomplish remarkable results. We can
plug into that nerve and thereby activate appropriate nerves
within the brain and achieve, at least to some degree, the
desired outcome. But we still don't know how to make appropriate direct connections to the brain ourselves without

the use of those existing input sensory nerves. We don't really even understand the basics of key brain functions, like memory. It is therefore unlikely that in the foreseeable future it will be possible to create a plug-in memory chip to augment human memory. We will indeed create microscopic computer memory chips with enormous capacity, but we just have no idea how to connect such things to the brain. There is no such thing as a "memory nerve" to make use of.

Humans cannot and will not be able to compete with machines in the area of memory. And one wonders if as advances in artificial intelligence continue we will be able to stay ahead of the machines in this critical area. Or is the purpose of humanity simply to usher in the era of the machines? Is our role only to serve as the creators for the next dominant form of intelligence on earth?

During the long history of life on earth it has always been the case that each species must evolve or perish. There is no species on the planet today, with very few exceptions, that much resembles its ancestors of 100 million years ago. Change is the norm. Adapt or die. Today is no different. And humans are no different. What is extraordinary is that for the first time in the history of the planet we now have a species able to control its own evolution, in terms of both direction and speed.

17

THE FUTURE

We face a future of incredible changes. For one thing, it will be possible to eliminate the suffering of genetic disease. Consider the single case of David Vetter. He was born September 21, 1971, in Houston, Texas, and received special treatment from the moment of his birth. His parents had a previous son, also named David, who had died at the age of seven months from a rare genetic disease called severe combined immunodeficiency disease, or SCID. This disease is the result of defective genes required for making the immune system. People with SCID have faulty immune systems and are extremely susceptible to infections, generally dying during their first year.

When the Vetters found out that they were expecting again they were told that another son would have a 50 percent chance of having SCID. And the only available treatment for SCID was a bone-marrow transplant from a well-matched donor. This would restore the stem cells that make the immune system, but in the 1970s the transplant would only succeed if the donor was a near-perfect match.

The doctors, having lost a series of children with SCID, were determined to do everything possible to save the Vetter child. They proposed keeping the baby in a sterile environment, protecting it from infections, while they tested to see if it had SCID. Then, if necessary, they would perform a bone-marrow transplant, or perhaps find another way to activate the immune system. The Vetters also had a healthy daughter, Katherine, and it was hoped that she would be a suitable match for the transplant. The sterile environment was intended to be a stopgap measure, to buy time while the baby was treated.

So David was delivered by cesarean section and immediately transferred to a sterile isolation chamber, a germ-free cocoon. The results of the genetic testing came about a week later; unfortunately, David did indeed have SCID. And further testing showed that Katherine was not very well matched with David after all. In addition, exhaustive searches of registries of bone-marrow-transplant donors in both the United Sates and Europe found no suitable match. David therefore seemed doomed to life in a sterile bubble.

At first David flourished. Indeed, as long as he was within his sterile bubble he never got sick. As David grew and crawled, and then walked, the size of his chamber was expanded, and a playroom was added on. But over the years the perpetual confinement led to a mental decline. David was often unhappy. The nurse would find David repeating, "One, two, three, four, I can't stand this anymore." As he got older the situation became more and more unbearable. David became preoccupied with death. His anxieties were getting worse and worse. People were concerned that in time he wouldn't be in command of his own behavior. He was having unmanageable crying episodes, and a psychologist diagnosed him as borderline psychotic.

Then there was a possible breakthrough, when researchers discovered a method for making bone-marrow transplants work with less-than-perfect matches. It wasn't thoroughly tested, but it did offer a glimmer of hope. It was decided to go ahead with a transplant, using Katherine's bone marrow. When he was twelve years old, on October 21, 1983, some of Katherine's bone marrow was transferred to David. There was great optimism. Everything seemed to go well for the first few months. The parents hoped that they would finally be able to hold their son in their arms.

But then his temperature shot up, his intestines started bleeding, and he vomited blood. The transplant wasn't working after all. In order to treat David it was necessary to take him from the bubble. Despite their best efforts, however, his condition deteriorated, and he lapsed into a coma. His parents were with him near the end. His mother asked if she could remove her glove, and the doctor nodded yes. Then, for the first and last time, she touched her son, caressing the back of his hand. Later that night he was pronounced dead.

An autopsy showed that a virus had contaminated the blood from Katherine. In the immune-deficient David it had caused a cancer, Burkitt's lymphoma, which had rapidly progressed and resulted in his death. It was one of the many scientific advances to come from David, a clear demonstration that in some cases viruses could cause cancer in people.

Today there are multiple treatment options for SCID patients. Transplant technology has advanced, making it possible to save patients using donors that are not perfect matches. In addition, in some cases it is possible to treat SCID patients using a method called enzyme-replacement therapy. One cause of SCID is a defective gene that encodes the enzyme adenosine deaminase (ADA). Without this enzyme there is a buildup of toxic chemicals in the cells that make up the immune system, killing them. It is possible to provide patients with the ADA enzyme, in a form that allows the cells to take it up, and to thereby reduce the levels of toxins and allow formation of the cells of the immune system. And now there is yet a third method of treatment of SCID patients, which is theoretically the most appealing of all, gene therapy.

GENE THERAPY

The basic concept of gene therapy is very simple. Many genetic diseases are the result of a defective gene in a critical type of cell. For example, as mentioned above, some SCID patients do not have a functional ADA gene in the cells that form their immune system. If we could just somehow insert a good ADA gene into these cells then they would be cured. The cells would survive, the immune system would be made normally, and the patients would be permanently healed.

The idea of gene therapy has been around since the late 1960s, but turning the idea into reality has been exceedingly difficult. How do you get the necessary genes into the right cells? Separating out the appropriate cells, the ones that need the good genes, is very challenging. It has taken many years to figure out how to isolate, for example, the stem cells that form the immune system. And then, how do you put the DNA into the cells? Some of the early efforts used a microinjection procedure. In essence, each cell was given a little shot of the DNA with the required good gene. But giving each cell a shot with a tiny needle, under a microscope, turned out to be very laborious as well as inefficient, because the injected DNA was usually lost over time, with only a very few of the injected cells actually keeping the gene and using it. Microinjection didn't really work.

The pioneers of gene therapy then turned to viruses, which were already masters of inserting their own genes into cells. The virus was simply genetically modified to include a good version of the gene that was defective in the patient, and then the virus was used as a little biological syringe to place the gene in the appropriate cells. This turned out to be a much better strategy than microinjection, but again there were many pitfalls.

The story of the development of gene therapy is filled with tragedy. We learned through trial and error that it was possible for the patient to die from a severe inflammatory reaction to the virus. And in several cases the virus, meant to save the patient by bringing in good genes, actually killed the patient by causing cancer. In addition there were many instances when the gene therapy did not harm the patient, but provided no benefit. Often the genes just didn't make their way into the right cells, even using the virus. And even when the virus did introduce the genes into the right cells, the genes usually did not stay active, or useful, for an extended period. For some reason, still poorly understood, the added genes turned off with time.

Dr. French Anderson was a leader in the field of gene therapy. Acknowledged by many as a genius, he was totally dedicated to making gene therapy work. Books have been written about him, including W. *French Anderson: Father of Gene Therapy,* by Bob Burke and Barry Epperson. After graduating from Harvard Medical School he took a job at the National Institutes of Health, and began gene-therapy research. In the year 1990, after many years of setbacks and struggles, he triumphantly claimed to have used gene therapy to successfully treat a young girl with SCID. He was a world-renowned hero. He presented his achievements to packed lecture halls, saving for the last slide a picture of himself walking with the young patient, holding her hand, and explaining how touching her was only possible because she had been cured. This was invariably followed by a rousing ovation. He received dozens of prestigious awards, and was runner-up for *Time* magazine's Person of the Year in 1995. But then it all came tumbling down. There were questions about the claimed gene-therapy cure. The SCID girl was also receiving enzyme-replacement therapy, and maybe this accounted for her improvement, and not the gene therapy. And then it got worse, when he was accused of molesting the young daughter of a co-worker. At the trial the prosecutors presented a secretly taped conversation between the wired accuser and Anderson, in which he appeared to admit guilt. He said, "I just did it, just something in me was evil." He was convicted, and sentenced to spend what will likely be the rest of his life in prison.

But the field of gene therapy survived its many problems, and researchers continued to gradually improve its technology. Different types of viruses were tested, and improved, making them more effective as delivery vehicles, and less harmful. In addition, new methods were developed to target the virus to the right cells, in some cases involving better purification of the cells to be genetically modified. And finally, forty years after the original concepts of gene therapy were developed, things seem to be coming together.

Katlyn DeMerchant, like David Vetter, was diagnosed with SCID at an early age. But, unlike David, Katlyn was not destined to spend her life in a sterile bubble. On May 26, 2007, Dr. Fabio Candotti and his team treated her with gene therapy at the National Institutes of Health, where French Anderson had worked. Stem cells that make the immune system were removed from her bone marrow and treated with a virus that injected into them a good copy of the ADA gene. The cells were returned to Katlyn, where they proceeded to make the immune system she had been lacking. Her family posted a blog of her progress at http://katlyn-adascid.blogspot.com/. While her immune system, even now, several years after the treatment, is not completely normal, Katlyn is nevertheless now able to lead a much more conventional life, including playing in the dirt, taking airplane trips, and having fun with her friends. To quote her mother, "She literally experiences new things and places on a daily basis. She has been borrowing books from our library as well as going fishing with her father. She spends pretty much her whole day outside enjoying every second of it. We were happy to finally have the yard to be able to get Katlyn a swing set. She loves it. Also we plan on putting Katlyn in swimming lessons this summer and maybe gymnastics in the fall. We are truly loving every second of life."

Katlyn isn't the only one to benefit from gene therapy. Many children with SCID have now been successfully treated, and other genetic diseases are beginning to yield. For example, Leber congenital amaurosis (LCA) causes severe visual loss in children. The cause is a defective *RPE65* gene, which is required for the photoreceptors of the retina of the eye to work properly and detect light. It is now possible to treat patients with a virus that will give the cells of the retina a functional *RPE65* gene. The procedure has been tried in a few people, and the results have been encouraging, with patients such as Corey Haas showing significant improvement in vision. Before the gene-therapy treatment, Corey was legally blind and used a cane and school materials in Braille. But one year after treatment, at age nine, he was able to walk in the woods, play baseball, drive go-carts, and read the blackboard in class. Many other genetic diseases, including hemophilia, Fanconi anemia, cystic fibrosis, and muscular dystrophy, might be treated with gene therapy in the future.

DESIGNER GENES

The gene-therapy story provides a reality check for those thinking about "designer-gene" children. Things that look easy in theory can sometimes take decades to actually happen. Unforeseen problems can crop up. Technologies that are advancing rapidly might bog down. For example, experts are predicting that within a few years it will be possible to completely sequence a person's DNA for under a thousand dollars. But maybe it will take much longer or cost more. And associating DNA sequences with certain traits might turn out to be more difficult than we currently think. One thing is certain, though, and that is the inexorable march forward that science makes. Any problems that arise will be overcome in time.

Another lesson from the gene-therapy story is that it is exceedingly difficult to correct defective genes in people.

Benjamin Franklin once said, "An ounce of prevention is worth a pound of cure." Why not, then, simply make sure that our children don't have defective genes in the first place? Today, in every children's hospital around the world, there are children confined to wheelchairs who will never be able to walk. And worse, there are the children with a blank stare who will never recognize their parents, and will never speak a word. There are a host of genetic conditions causing everything from very low IQ to muscles that don't work, nerves that don't work, blood cells that don't work, enzymes that don't work, and, in the end, children that don't work. One must admire and respect the parents who provide the care and unending love for these children. But this is not a perfect world.

A more widespread use of genetic screening could prevent many of these unfortunate situations. It is now possible to test for the presence of over a hundred genetic diseases, and this number will continue to grow as we sequence the DNA of more people and associate more genes with disease. One future approach might be to simply analyze the DNA of most or all embryos, using chorionic villus sampling or amniocentesis, equipped with our increasing knowledge of the consequences of different DNA sequences. But, as the bumper stickers remind us, these later-stage embryos do have a beating heart. Could we choose to abort such a fetus just because it was genetically destined to have a slight facial deformity, or might suffer a mild genetic disease late in life, or would have a less than stellar IQ? There is an obvious relationship between the stage of the pregnancy and the degree of allowed selection in the screening process. If a baby is about to be born then there is probably no DNA information that would convince us to abort it, but if we are dealing with an embryo that is but a tiny clump of a few cells then the situation is very different.

GENETICALLY ENGINEERED EGGS AND SPERM

There is one option that allows all of the benefits of genetic engineering and trait selection without any of the moral objections associated with the destruction of embryos. It pushes the rule of "earlier is better" to the extreme. What if we genetically engineered the eggs and the sperm instead of the embryos? Could we somehow make the ideal egg, and the ideal sperm, and then fuse them together to make the ideal embryo? In theory at least this might indeed be possible, and without the sacrifice of any embryos. In essence, stem cells made from the parents would be used to make designer genes eggs and sperm.

The strategy would be to first take adult cells from the prospective mother and father, perhaps from the skin, and turn them into stem cells, using the methods described earlier. These stem cells could then be genetically optimized using gene-targeting technology. The stem cells from the mother would then be made into eggs, and the stem cells from the father into sperm, using methods that are still being developed. Many papers have been published on the topic of turning stem cells into eggs and sperm, and much progress has been made, but we have not yet completely figured out how to do it. This technical hurdle would have to be overcome before this strategy could be made to work.

The individual eggs and sperm would then be subjected to DNA-sequence analysis to find those with the best combinations of genes. This might sound easy, but it would actually be difficult, because you can't sequence the DNA of a cell without killing it. So it would be necessary to use a method of subtractive sequence analysis. The DNA sequence of the starting stem cells would be determined. These cells, of course, have two copies of every gene. When eggs and sperm are made, either natually or from stem cells, there is a required "reduction" cell division, which reduces the number of copies of each gene from two to one. One daughter cell receives one copy of each gene, with the other daughter cell getting the other copy. If we sequence the DNA of one daughter cell, then we can deduce the sequence of the other, without killing it.

The selected sperm would then be used to fertilize the preferred egg, perhaps using a process called zona drilling, which is sometimes employed to help infertile couples conceive. Micro instruments are used, under a microscope, to drill a hole through the shell of the egg and to insert the sperm, thereby achieving fertilization. The resulting embryo would then be grown for a few days in the laboratory, to make sure all was well, and then introduced into the uterus of the mother, where it would develop into a normal baby. And no embryos would be harmed in the process!

While this genetic engineering of egg and sperm might be the most appealing procedure, it is also the most technically challenging, and cannot be achieved at present. The Hinxton Group, an international consortium of experts on stem cells, ethics, and the law, stated in April 2008, "The derivation of human eggs and sperm in vitro from PSCs"—pluripotent stem cells (stem cells made from adults)—"in whole or at least in part, is anticipated within 5 to 15 years." That is, they predicted that it would become possible to take stem cells (perhaps made from adult skin), and to then use those stem cells to make eggs and sperm, perhaps as early as 2013, but most likely at least by 2023. This projection, however, might have been too conservative. Already in the year 2009 the laboratory of Dr. Karim Nayernia reported using human embryonic stem cells to make sperm that appeared normal in most respects. They had one copy of each gene instead of two, they could swim, and they had the proteins necessary to activate the egg upon fertilization. We are clearly making rapid progress in this area.

In the meantime, it is possible right now to remove single cells from early embryos, to analyze their DNA, and to predict certain traits.

THE FUTURE IS NOW

A former biopharmaceutical company named deCODE genetics used DNA analysis techniques, combined with medical records from hundreds of thousands of volunteers, to associate certain genes with different diseases and physical traits. In 2007 it published a paper in the very prestigious journal *Nature Genetics* describing various DNA sequences associated with several distinct physical traits. To quote the paper, "The variants described in this report enable prediction of pigmentation traits based upon an individual's DNA." Using deCODE's results, and others, it is now possible to look at DNA sequence and to make predictions concerning eye color, hair color, height, complexion, and body build.

Dr. Jeffrey Steinberg, the head of Fertility Institutes, has proposed making trait-related DNA sequence information available to customers during the early-embryo screening process. It will then be possible to choose not only a healthy gene combination, ensuring the absence of known genetic diseases, but also the attributes of pigmentation and body build. Some refer to this as cosmetic genetics. If we are willing to use cosmetics, dye our hair, and work out to improve our appearance, then why not choose gene combinations that would make our children attractive? According to the American Society for Aesthetic Plastic Surgery there were 11.9 million procedures performed in the United States alone in 2004. While appearance is not a functional trait, like intelligence and athletic ability, it is nevertheless clearly considered important by many. And although the list of selectable traits is relatively short right now, in the future it will only expand to include more and more features.

We will have the ability to choose the genes of our children, but will we do it? Just because we can do something doesn't mean that we will. Nevertheless, there are many reasons to think that if given the opportunity to genetically enhance their children, people would indeed do it. Our children are our love. If they are sick we will spare no expense to make them well. We send them to the best schools we can possibly afford to enhance their intellectual development. We will make any sacrifice, spare no expense, to give them every opportunity. Our children are our future. It would seem a natural extension of this desire to help our children to want to give them the best genes possible.

To quote James Watson, "Once you have a way in which you can improve our children, no one can stop it. It would be stupid not to use it because someone else will. Those parents who enhance their children, then their children are going to be the ones who dominate the world." Watson goes on to discuss the idea that traits like beauty could also be genetically engineered. "People say it would be terrible if we made all girls pretty. I think it would be great."

The genetics revolution is ongoing, and there may indeed be no stopping it now. The human species is about to undergo an incredible transformation. The forces of natural selection are about to be replaced by the forces of human selection. In the future it will likely be routine for parents to choose the genes of their children. Where this might take us we cannot know. Of course we will quickly remove from the human population versions of genes that result in catastrophic disease. But in addition we will be selecting ideal combinations of genes from the two parents, modifying genes for which neither parent offers a preferred copy, and perhaps even making entirely new kinds of genes never before seen in nature. For the first time humans will have complete control over their genetic destiny. And the process will result in an upward spiral of genetic change, as each generation is more intelligent than the last and better able to choose the genetic makeup of their children.

It could well mean the end of the human race as we know it, but perhaps the beginning of something better.

INDEX

STEVEN POTTER was an undergraduate at UCLA, received his Ph.D. from the University of North Carolina at Chapel Hill, and was a Postdoctoral Research Fellow at Harvard Medical School. He is currently a professor in the department of pediatrics, division of developmental biology, at Children's Hospital Medical Center in Cincinnati. Potter has published over a hundred research papers, including more than a dozen in the prestigious journals *Nature, Cell,* and *Science.* He also co-authored the third edition of the medical school textbook *Larsen's Human Embryology,* and serves on the editorial boards of the science journals *Transgenics* and *Developmental Biology.* His research has covered a wide range of topics, including evolution, jumping genes, targeted modification of genes, and the study of how organs form. In his spare time he likes to ride his motorcycle with his son and go backpacking in the U.S. Southwest with his daughter.

ABOUT THE TYPE

This book was set in Sabon, a typeface designed by the well-known German typographer Jan Tschichold (1902–74). Sabon's design is based upon the original letter forms of Claude Garamond and was created specifically to be used for three sources: foundry type for hand composition, Linotype, and Monotype. Tschichold named his typeface for the famous Frankfurt typefounder Jacques Sabon, who died in 1580.